The Watershed

ARTHUR KOESTLER, born in Budapest in 1905 and educated in Vienna, began his writing career as an editor of a German and Arabian weekly newspaper in Cairo. This led to his becoming the Near East correspondent for the Ullstein newspapers of Berlin. Future assignments included the Graf Zepplin's flight over the Arctic in 1931 and the Civil War in Spain, where he was imprisoned, sentenced to death, and finally pardoned by the rebels. An active communist from 1931 to 1938, Koestler achieved world fame with DARKNESS AT NOON (1941), his explosive anti-communist novel. He now lives in England. A more comprehensive biography appears in John Durston's Foreword.

THE
WATERSHED

A Biography of Johannes Kepler

Arthur Koestler

With a Foreword by John Durston

UNIVERSITY
PRESS OF
AMERICA

University Press of America,™ Inc.

4720 Boston Way
Lanham, MD 20706

Library of Congress Cataloging in Publication Data

Koestler, Arthur, 1905-
    The watershed : a biography of Johannes Kepler.

    "The watershed is taken from . . . The sleepwalkers"—
Reprint. Originally published: Garden City, N.Y. :
Anchor Books, 1960. (Science study series ; S16)
    Bibliography: p.
    Includes index.
    1. Kepler, Johannes, 1571-1630.   2. Astronomers—
Germany—Biography.   I. Title.
QB36.K4K6   1985   520'.92'4 [B]   84-19690
ISBN 0-8191-4339-1 (pbk. : alk. paper)

Reprinted by arrangement with
Doubleday & Company, Inc.

Illustrated by R. Paul Larkin

THE SCIENCE STUDY SERIES

The Science Study Series offers to students and to the general public the writing of distinguished authors on the most stirring and fundamental topics of physics, from the smallest known particles to the whole universe. Some of the books tell of the role of physics in the world of man, his technology and civilization. Others are biographical in nature, telling the fascinating stories of the great discoverers and their discoveries. All the authors have been selected both for expertness in the fields they discuss and for ability to communicate their special knowledge and their own views in an interesting way. The primary purpose of these books is to provide a survey of physics within the grasp of the young student or the layman. Many of the books, it is hoped, will encourage the reader to make his own investigations of natural phenomena.

These books are published as part of a fresh approach to the teaching and study of physics. At the Massachusetts Institute of Technology during 1956 a group of physicists, high school teachers, journalists, apparatus designers, film producers, and other specialists organized the Physical Science Study Committee, now operating as a part of Educational Services Incorporated, Watertown, Massachusetts. They pooled their knowledge and

The Science Study Series

experience toward the design and creation of aids to the learning of physics. Initially their effort was supported by the National Science Foundation, which has continued to aid the program. The Ford Foundation, the Fund for the Advancement of Education, and the Alfred P. Sloan Foundation have also given support. The Committee is creating a textbook, an extensive film series, a laboratory guide, especially designed apparatus, and a teachers' source book for a new integrated secondary school physics program which is undergoing continuous evaluation with secondary school teachers.

The Series is guided by a Board of Editors, consisting of Paul F. Brandwein, the Conservation Foundation and Harcourt, Brace and Company; John H. Durston, Educational Services Incorporated; Francis L. Friedman, Massachusetts Institute of Technology; Samuel A. Goudsmit, Brookhaven National Laboratory; Bruce F. Kingsbury, Educational Services Incorporated; Philippe LeCorbeiller, Harvard University; and Gerard Piel, *Scientific American*.

8

FOREWORD

The Watershed is taken from Arthur Koestler's *The Sleepwalkers* (The Macmillan Company, 1959), which bears the subtitle, "A History of Man's Changing Vision of the Universe." The Board of Editors has included it in the Science Study Series because its subject, Johannes Kepler, is a significant figure in the history of science and because it is the first essay in any language to explore Kepler's weird genius. Over the centuries science has made great use of Kepler's works but ignored the man. Few of his contemporaries tried, or cared, to understand him, and his famous laws, resurrected by Isaac Newton, were all of him that seemed to have escaped the grave. But time, happily, has a way of redressing balances. In this biography, for which the world had to wait three hundred and thirty years, Kepler the man emerges as a rich, if outlandish, character, deserving of immortality.

To do any genius justice takes a practiced pen, and if he is outlandish to boot, the task is magnified a hundred times over. It is fortunate that Kepler should have caught the fancy of an accomplished novelist like Mr. Koestler, and it is doubly fortunate that Mr. Koestler should have a more than passing acquaintance with physics. True, he has made his reputation outside the field of science—his novel *Darkness at Noon* was one of the genuinely influential books of our mid-century—but

9

physical science was an early love, to which, in recent years, he has returned.

Mr. Koestler studied physics at the engineering school of the University of Vienna in the 1920s and later earned his living writing about science for lay readers. A Hungarian by birth, he might be described as a European-at-large; since boyhood he has spent little time in his native land. When he left the engineering school, he roamed the Near East for a while, farmed in Palestine, hawked lemonade about the streets of Haifa, and edited a German and Arabian weekly newspaper in Cairo. This last occupation led to his becoming Near East correspondent for the Ullstein newspapers, a publishing house of great respectability in pre-World War II Berlin.

His journalistic career broadened his horizons. As Ullstein science editor, he met some of the outstanding men of twentieth-century physics. He went to the Arctic in the Graf Zeppelin, and he visited Samarkand and other remote places of Soviet Central Asia. He was a free-lance correspondent in Paris. In the Spanish Civil War he was a correspondent for the London *News Chronicle*; the rebels captured him and would have executed him if the British Foreign Office had not intervened. In World War II he spent weary months in various alien detention camps, where his main concern was to keep out of the hands of Nazi SS troopers, and he served in the French and British armies. An ardent Zionist since youth, he returned to Palestine after the war. All these adventures and experiences you can read about in his books, including his autobiography. He now makes his home in England.

Mr. Koestler was an active Communist from 1931 to 1938. Firsthand observation of the deceits, the brutalities, and the inhumanity of Stalin's brand of Communism disillusioned him and awakened him to its threat to civilization. This disillusionment resulted in *Dark-*

ness at Noon, a powerful psychological study of an old Bolshevik victim of the Soviet purge trials of the 1930s. It is not too much to say that this book was a major influence in opening the eyes of the intellectuals of Western Europe and the United States to the perils of Communism.

While war, Communism, and Zionism seem far afield from elliptical orbits, Mr. Koestler's experiences among the persecuted and displaced of this world have given him an understanding of human oddities that enlivens and enlarges his portrait of Kepler. It is his thesis in *The Watershed* that this strange, tormented genius stood astride the historical crest—the intellectual divide —that separated ancient and medieval thought from modern observational science — hence the title. For young people interested in physics—and the Science Study Series is dedicated to them—*The Watershed* is an incomparable account of one of the momentous discoveries. Neither in his approach to science nor in his approach to life is Kepler to be recommended as an ideal pattern, but in this biography there is much to learn about history, physics, and people.

JOHN H. DURSTON

CONTENTS

Contents

Contents

chapter one

The Young Kepler

1. DECLINE OF A FAMILY

Johannes Kepler, Keppler, Khepler, Kheppler, or Keplerus, the founder of modern astronomy, was conceived on May 16, A.D. 1571, at 4:37 A.M., and was born on December 27 at 2:30 P.M., after a pregnancy lasting 224 days, 9 hours, and 53 minutes. The five different ways of spelling his name are all his own, and so are the figures relating to conception, pregnancy, and birth, recorded in a horoscope which he cast for himself.[1] The contrast between his carelessness about his name and his extreme precision about dates reflects, from the very outset, a mind to which all ultimate reality, the essence of religion, of truth and beauty, was contained in the language of numbers.

He was born in the township of Weil in wine-happy Swabia, a blessed corner of southwest Germany between the Black Forest, the Neckar, and the Rhine. Weil-der-Stadt—a freak name, meaning Weil-the-Town, but with the masculine *der* instead of the feminine *die*—has beautifully succeeded in preserving its medieval character to our day.* It stretches along the top of a mound, long and narrow, like the hull of a battleship, surrounded by massive, crenelated, ocher-colored walls, and slender watch-

* At least, to be precise, to the days of May 1955, when I visited Kepler's birthplace.

towers topped by spire and weathercock. The gabled houses, with their irregular patterns of small, square windows, are covered with scarab-green, topaz-blue, and lemon-yellow stucco on their cockeyed façades; where the stucco peels, the mud and lath peep through like weathered skin showing through a hole in a peasant's shirt. If, after fruitless knocking, you push open the door of a house, you are likely to be greeted by a calf or a goat, for the ground floors of some old houses still serve as stables, with an inner staircase leading up to the family's living quarters. The warm smell of compost floats everywhere in the cobbled streets, but they are kept scrupulously, Teutonically clean. The people speak a broad Swabian dialect and frequently address even the stranger with "thou"; they are rustic and *gemuetlich*, but also alert and bright. There are places outside the walls still called "God's Acre" and "Gallows Hill"; and the old family names, down from the mayor, Herr Oberdorfer, to the watchmaker, Herr Speidel, are the same that appear on documents from Kepler's time, when Weil had only two hundred citizens. Though it produced some other distinguished men—among them the phrenologist Gall, who traced each faculty of the mind to a bump on the skull—Johannes Kepler is the town's hero, venerated like a patron saint.[2]

One of the entries, dated 1554, in the municipal ledger, refers to the lease of a cabbage patch to Johannes' grandfather, Sebaldus Kepler.

> "Daniel Datter and Sebold Kepler, furrier, shall pay seventeen pennies at Martinmas out of their cabbage patch on the Klingelbrunner Lane between the fields of Joerg Rechten and those of Hans Rieger's children. Should they relinquish the cabbage patch, they shall cart six cartloads of compost into or onto it."

From this bucolic prelude one would expect a happy

childhood for the infant Johannes. It was a ghastly one.

Grandfather Sebaldus, the furrier with the cabbage patch, was said to stem from a noble family,[3] and became mayor of Weil; but after him, the respectable Keplers went into decline. His offspring were mostly degenerates and psychopaths, who chose mates of the same ilk. Johannes Kepler's father was a mercenary adventurer who narrowly escaped the gallows. His mother, Katherine, an innkeeper's daughter, was brought up by an aunt who was burned alive as a witch, and Katherine herself, accused in old age of consorting with the Devil, had as narrow an escape from the stake as the father had from the gallows.

Grandfather Sebaldus' house (burned down in 1648 but rebuilt later in the same style) stood on a corner of the market place. Facing the house is a beautiful Renaissance fountain with four long, fluted copper spouts which issue from four human faces carved into the stone. Three of the faces are stylized masks; the fourth, turned toward the town hall and the Kepler house, looks like the realistic portrait of a bloated, coarse-featured man. There is a tradition in Weil according to which it is the likeness of old Sebaldus, the mayor. This may or may not be so, but it fits Kepler's own description of him.

"My grandfather Sebald, mayor of the imperial city of Weil, born in the year 1521 about St. James's day . . . is now 75 years of age. . . . He is remarkably arrogant and proudly dressed . . . short-tempered and obstinate, and his face betrays his licentious past. It is a red and fleshy face, and his beard gives it much authority. He was eloquent, at least as far as an ignorant man can be. . . . From the year 1578 onward his reputation began to decline, together with his substance. . . ."[4]

This thumbnail sketch, and the others which follow, are part of a kind of genealogical horoscope, embracing all members of his family (including himself), which Kepler drew up when he was twenty-six. It is not only a remarkable document, but also a precious contribution to the study of the hereditary background of genius, for it happens only rarely that the historian has such ample material at his disposal.*

When Grandfather Sebald was twenty-nine, he married Katherine Mueller, from the nearby village of Marbach. Kepler describes her as:

"restless, clever, and lying, but devoted to religion; slim and of a fiery nature; vivacious, an inveterate troublemaker; jealous, extreme in her hatreds, violent, a bearer of grudges. . . . And all her children have something of this. . . ."[5]

He also accuses his grandmother of pretending that she married at eighteen, when she was really twenty-two. However that may be, she bore Sebaldus twelve children in twenty-one years. The first three, named Sebaldus, Johan, and Sebaldus, all died in infancy. The fourth was Kepler's father, Heinrich, whom we leave aside for a moment. Of numbers 5 to 9 among his aunts and uncles, Kepler records:[6]

"5. Kunigund, born 1549, 23 May. The moon could not have been worse placed. She is dead, the mother of many children, poisoned, they think, in the year 1581, 17 July" [added later on: "Otherwise she was pious and wise"].†

"6. Katherine, born 1551, 30 July. She too is dead.

* As the document is a horoscope, events and character traits are derived from planetary constellations, which I have mostly left out.

† In later years, Kepler added a few remarks to his text, which soften, and sometimes contradict, the trenchant characterizations of his youth. I have put these addenda into brackets.

"7. Sebaldus, born 1552, 13 November.* An astrologer and a Jesuit, he underwent the first and second ordinations for the priesthood; though a Catholic, he imitated the Lutherans and led a most impure life. Died in the end of dropsy after many earlier illnesses. Acquired a wife who was rich and nobly born, but one of many children. . . . Was vicious and disliked by his fellow townsmen. In 1576, 16 August, he left Weil for Speyr, where he arrived on the 18th; on the 22 December he left Speyr against the will of his superior and wandered in extreme poverty through France and Italy. [He was held to be kind and a good friend.]

"8. Katherine, born 1554, 5 August. She was intelligent and skillful, but married most unfortunately, lived sumptuously, squandered her goods, now a beggar. [Died in 1619 or 1620.]

"9. Maria, born 1556, 25 August. She too is dead."

Of Numbers 10 and 11, he has nothing to say; Number 12, the last-born of his uncles and aunts, also died in infancy.†

All these misshapen progeny—except those who died in their cots—lived with old choleric Sebaldus and his shrewish wife, crowded into the narrow Kepler house, which, in fact, was rather a cottage. Kepler's father, Heinrich, though the fourth child, was the oldest among those who survived, and thus inherited the house, pro-

---

* The grandparents' third and last attempt to produce a Sebaldus who would survive.

† Cf. Kretschmer: "One is tempted to say: genius arises in the hereditary process particularly at that point where a highly gifted family begins to degenerate. . . . This degeneration often announces itself in the generation to which the genius belongs, or even in the preceding one, and generally in the form of psychopathic and psychotic conditions."[6a]

ducing seven children in his turn. Kepler describes him thus:

> "4. Henrich, my father, born 1547, 19 January.
> . . . A man vicious, inflexible, quarrelsome, and doomed to a bad end. Venus and Mars increased his malice. Jupiter combust[7] in descension made him a pauper but gave him a rich wife. Saturn in VII made him study gunnery; many enemies, a quarrelsome marriage . . . a vain love of honors, and vain hopes about them; a wanderer. . . . 1577: he ran the risk of hanging. He sold his house and started a tavern. 1578: a hard jar of gunpowder burst and lacerated my father's face. . . . 1589: treated my mother extremely ill, went finally into exile and died."

There is not even the usual mitigating addendum at the end. The story behind the entries is briefly this.

Heinrich Kepler married at the age of twenty-four. He seems to have studied no trade or craft, except "gunnery," which refers to his later military adventures. Seven months and two weeks after his marriage to Katherine Guldenmann, Johannes Kepler was born. Three years later, after the birth of his second son, Heinrich took the Emperor's shilling and went off to fight the Protestant insurgents in the Netherlands—an act the more ignominious as the Keplers were among the oldest Protestant families in Weil. The next year, Katherine joined her husband, leaving her children in the care of the grandparents. The year after, they both returned, but not to Weil, where they were disgraced; instead, Heinrich bought a house in nearby Leonberg; but in a short time left again for Holland, to join the mercenary hordes of the Duke of Alba. It was apparently on this journey that he "ran the risk of hanging" for some unrecorded crime. He returned once more, sold the house in Leonberg, ran a tavern in Ellmendingen, again went back to Leonberg, and in 1588 vanished forever from the sight of his family. Rumor had it that he enlisted in the Neapolitan fleet.

His wife Katherine, the innkeeper's daughter, was an equally unstable character. In the family horoscope, Kepler describes her as: "small, thin, swarthy, gossiping, and quarrelsome, of a bad disposition." There was not much to choose between the two Katherines, the mother and the grandmother; and yet the mother was the more frightening of the two, with an aura of magic and witchcraft about her. She collected herbs and concocted potions in whose powers she believed; I have already mentioned that the aunt who brought her up had ended her days at the stake, and that Katherine nearly shared the same fate, as we shall hear.

To complete the survey of this idyllic family, I must mention our Johannes' brothers and sisters. There were six of them; of whom three again died in childhood, and two became normal, law-abiding citizens (Gretchen, who married a vicar, and Christopher, who became a pewterer). But Heinrich, the next in age to Johannes, was an epileptic and a victim of the psychopathic streak running through the family. An exasperating problem child, his youth seems to have been a long succession of beatings, misadventures, and illnesses. He was bitten by animals, nearly drowned, and nearly burned alive. He was apprenticed to a draper, then a baker, and finally ran way from home when his loving father threatened to sell him. In subsequent years, he was a camp follower with the Hungarian army in the Turkish wars, a street singer, baker, nobleman's valet, beggar, regimental drummer, and halberdier. Throughout this checkered career, he remained the hapless victim of one misadventure after another—always ill, sacked from every job, robbed by thieves, beaten up by highwaymen—until he finally gave up, begged his way home to his mother, and hung to her apron strings until he died at forty-two. In

23

his childhood and youth, Johannes conspicuously shared some of his younger brother's attributes, particularly his grotesque accident-proneness and constant ill health combined with hypochondria.

2. JOB

Johannes was a sickly child, with thin limbs and a large, pasty face surrounded by dark curly hair. He was born with defective eyesight—myopia plus anocular polyopy (multiple vision). His stomach and gall bladder gave constant trouble; he suffered from boils, rashes, and possibly from piles, for he tells us that he could never sit still for any length of time and had to walk up and down.

The gabled house on the market place in Weil, with its crooked beams and doll-house windows, must have been bedlam. The bullying of red-faced old Sebaldus; the high-pitched quarrels of mother Katherine and grandmother Katherine; the brutality of the weak-headed, swashbuckling father; the epileptic fits of brother Heinrich; the dozen or more seedy uncles and aunts, parents and grandparents, all crowded together in that unhappy little house.

Johannes was four years old when his mother followed her husband to the wars; five, when the parents returned and the family began its restless wanderings to Leonberg, Ellmendingen, and back to Leonberg. He could attend school only irregularly, and from his ninth to his eleventh year did not go to school at all but was "put to hard work in the country." As a result, and in spite of his precocious brilliance, it took him twice as long as it took normal children to complete the three classes of the elementary Latin school. At thirteen, he was at last able to enter the lower theological seminary at Adelberg.

The notes on his own childhood and youth, in the family horoscope, read like the diary of Job.

The Young Kepler

"On the birth of Johann Kepler. I have investigated the matter of my conception, which took place in the year 1571, May 16, at 4:37 A.M. . . . My weakness at birth removes the suspicion that my mother was already pregnant at the marriage, which was the 15th of May. . . . Thus I was born premature, at thirty-two weeks, after 224 days, ten hours. . . .1575 [aged four].I almost died of smallpox, was in very ill health, and my hands were badly crippled. . . . 1577 [aged six]. On my birthday I lost a tooth, breaking it off with a string which I pulled with my hands. . . . 1585–86 [fourteen–fifteen]. During these two years, I suffered continually from skin ailments, often severe sores, often from the scabs of chronic putrid wounds in my feet which healed badly and kept breaking out again. On the middle finger of my right hand I had a worm, on the left a huge sore. . . . 1587 [sixteen]. On April 4 I was attacked by a fever. . . . 1589 [nineteen]. I began to suffer terribly from headaches and a disturbance of my limbs. The mange took hold of me. . . . Then there was a dry disease. . . .1591 [twenty]. The cold brought on prolonged mange. . . . A disturbance of body and mind had set in because of the excitement of the Carnival play in which I was playing Mariamne. . . . 1592 [twenty-one]. I went down to Weil and lost a quarter florin at gambling. . . ."

Only two brief memories mitigate the gloom and squalor of this childhood. At the age of six:

"I heard much of the comet of that year, 1577, and was taken by my mother to a high place to look at it."

And at the age of nine:

"I was called outdoors by my parents especially, to

look at the eclipse of the moon. It appeared quite red."

So much for the sunny side of life.

No doubt, some of his miseries and ailments existed only in his imagination; while others—all these cold sores, worms on the finger, scabs, and manges—seem like the stigmata of his self-detestation, physical projections of the image he had formed of himself: the portrait of a child as a mangy dog. He meant this literally, as we shall see.

3. ORPHIC PURGE

There are always compensations. For Kepler, the compensations offered by destiny were the exceptional educational facilities in his native land.

The Dukes of Wuerttemberg, after embracing the Lutheran creed, had created a modern educational system. They needed erudite clergymen who could hold their own in the religious controversy that was raging across the country, and they needed an efficient administrative service. The Protestant universities in Wittenberg and Tuebingen were the intellectual arsenals of the new creed; the confiscated monasteries and convents provided ideal accommodation for a network of elementary and secondary schools, which fed the universities and chancelleries with bright young men. A system of scholarships and grants for "the children of the poor and faithful who are of a diligent, Christian, and god-fearing disposition" vouchsafed an efficient selection of candidates. In this respect, Wuerttemberg before the Thirty Years' War was a modern welfare state in miniature. Kepler's parents would certainly not have bothered about his education; the precocious brilliance of the child automatically guaranteed his progress from school to seminary and from there to university, as on a moving belt.

The curriculum at the seminary was in Latin, and the pupils were rigorously held to use only Latin even among themselves. Even in the elementary school, they were made to read the comedies of Plautus and Terence, to add colloquial fluency to scholarly precision. The German vernacular, though it had acquired a new dignity through Luther's Bible translation, was not yet considered a worthy medium of expression for scholars. As a happy result of this, Kepler's style, in those pamphlets and letters which he wrote in German, has an enchantingly naïve and earthy quality which, in contrast to the dehydrated medieval Latin, sounds like the joyous din of a country fair after the austerities of the lecture room. Kepler's German seems modeled on Luther's pronouncement: "One should not imitate those asses who ask the Latin language how German should be spoken; but should ask the mother in her home, the children in the gutters, the common man at the fair, and watch their big mouths as they speak, and do accordingly."

When he had passed the Elementary Latin school, Johannes' good brains, bad health, and interest in religion made the career of a clergyman the obvious choice. The theological seminary he attended from his thirteenth to his seventeenth year was divided into a lower (Adelberg) and a higher course (Maulbronn). The curriculum was broad and rounded, adding Greek to Latin, and embracing, besides theology, the study of the pagan classics, rhetoric and dialectics, mathematics and music. Discipline was strict: classes started in summer at four, in winter at five o'clock in the morning; the seminarists had to wear a sleeveless, shapeless cloak reaching below their knees, and were hardly ever allowed out on leave. Young Kepler recorded two of his most daring and paradoxical utterances from his seminarist days: that the study of philosophy was a symptom of

27

Germany's decline; and that the French language was worthier of study than the Greek. No wonder his fellows regarded him as an intolerable egghead and beat him up at every opportunity.

He was, indeed, as unpopular among his schoolmates as he was beloved by his friends in later years. In his horoscope record, the entries relating his physical afflictions alternate with others that reveal his moral misery and loneliness.

"February, 1586. I suffered dreadfully and nearly died of my troubles. The cause was my dishonor and the hatred of my school fellows whom I was driven by fear to denounce. . . . 1587. On April 4 I was attacked by a fever from which I recovered in time, but I was still suffering from the anger of my schoolmates with one of whom I had come to blows a month before. Koellin became my friend; I was beaten in a drunken quarrel by Rebstock; various quarrels with Koellin. . . . 1580. I was promoted to the rank of Bachelor. I had a most iniquitous witness, Mueller, and many enemies among my comrades. . . ."

The narrative of the horoscope was continued in the same year (his twenty-sixth) in another remarkable document, a self-analysis more unsparing than the *Confessions* of Jean Jacques Rousseau.[8] Written in the year when his first book was published, when he had undergone a kind of orphic purge and found his final vocation, it is perhaps the most introspective piece of writing of the Renaissance. Several pages of it describe his relations with colleagues and teachers at the seminary, and later at the University of Tuebingen. Referring to himself in the third person, as he mostly does in this document, the passage begins: "From the time of his arrival [at the seminary] some men were his adversaries." He lists five of them, then continues: "I record the

most lasting enemies." He lists another seventeen, "and
many other such." He explains their hostility mainly on
the grounds that "they were always rivals in worth, hon-
ors, and success." There follows a monotonous and depressing record of these enmities and quarrels. Here are
samples.

"Kolinus did not hate me, rather I hated him.
He started a friendship with me, but continually
opposed me. . . . My love of pleasure and other
habits turned Braunbaum from being a friend into
an equally great enemy. . . . I willingly incurred
the hatred of Seiffer because the rest hated him
too, and I provoked him although he had not
harmed me. Ortholphus hated me as I hated Kolinus, although I on the contrary liked Ortholphus,
but the rivalry between us was many-sided. . . . I
have often incensed everyone against me through
my own fault: at Adelberg it was my treachery [in
denouncing his schoolmates]; at Maulbronn, my
defense of Graeter; at Tuebingen, my violent request for silence. Lendlinus I alienated by foolish
writings, Spangenburg, by my temerity in correct-
ing him when he was my teacher; Kleberus hated
me as a rival. . . . The reputation of my talent
annoyed Rebstock, and also my frivolousness. . . .
Husalius opposed my progress. . . . With Dauber
there was a secret rivalry and jealousy. . . . My
friend Jaeger betrayed my trust: he lied to me and
squandered much of my money. I turned to hatred
and exercised it in angry letters during the course of
two years."

And so on. The list of friends turned into enemies
ends with the pathetic remark:

"Lastly, religion divided Crellius from me, but
he also broke faith; henceforth I was enraged with

him. God decreed that he should be the last. And so the cause was partly in me and partly in fate. On my part anger, intolerance of bores, an excessive love of annoying and of teasing, in short of checking presumptions. . . ."

Even more pathetic is the one exception in the list.

"Lorhard never communicated with me. I admired him, but he never knew this, nor did anyone else."

Immediately following this dismal recital, Kepler put down, with acid amusement, this portrait of himself—where the past tense alternates revealingly with the present.[9]

"That man [i.e., Kepler] has in every way a dog-like nature. His appearance is that of a little lap dog. His body is agile, wiry, and well proportioned. Even his appetites were alike: he liked gnawing bones and dry crusts of bread, and was so greedy that whatever his eyes chanced on he grabbed; yet, like a dog, he drinks little and is content with the simplest food. His habits were similar. He continually sought the good will of others, was dependent on others for everything, ministered to their wishes, never got angry when they reproved him, and was anxious to get back into their favor. He was constantly on the move, ferreting among the sciences, politics, and private affairs, including the lowest kind; always following someone else, and imitating his thoughts and actions. He is bored with conversation, but greets visitors just like a little dog; yet when the least thing is snatched away from him, he flares up and growls. He tenaciously persecutes wrongdoers—that is, he barks at them. He is malicious and bites people with his sarcasms. He hates many people exceedingly and they avoid him, but his masters are fond of him. He has a doglike horror of baths, tinctures, and lotions. His recklessness knows no limits, which is surely due to Mars in quadrature with Mercury, and in trine with the moon; yet he takes good care of his life. . . . [He has] a vast appetite for the greatest things. His teachers praised him for his good dispositions, though morally he was the worst among his contemporaries. . . . He was religious to the point of superstition. As a boy of ten years when he first read Holy Scripture . . . he grieved that on account of the impurity of his life, the honor to be a prophet was denied him. When he committed a wrong, he performed an expiatory rite, hoping it would save him from punishment: this consisted in reciting his faults in public. . . .

In this man there are two opposite tendencies: always to regret any wasted time, and always to waste it willingly. For Mercury makes one inclined to amusements, games, and other light pleasures. . . . Since his caution with money kept him away from play, he often played by himself. It must be noted that his miserliness did not aim at acquiring riches, but at removing his fear of poverty—although, perhaps avarice results from an excess of this fear. . . ."

Of love, there is an isolated obscure entry, referring to his twentieth year.

"1591. The cold brought on prolonged mange. When Venus went through the Seventh House, I was reconciled with Ortholphus: when she returned, I showed her to him; when she came back a third time, I still struggled on, wounded by love. The beginning of love: April 26."

31

That is all. We are told no more about that nameless "she."

We remember that Kepler wrote this at the age of twenty-six. It would be a harsh self-portrait even for a modern young man, reared in the age of psychiatry, anxiety, masochism, and the rest; coming from a young German at the close of the sixteenth century, the product of a coarse, brutal, and callow civilization, it is an astonishing document. It shows the ruthless intellectual honesty of a man whose childhood was spent in hell and who had fought his way out of it.

With all its rambling inconsequences, its baroque mixture of sophistication and naïveté, it unfolds the timeless case history of the neurotic child from a problem family, covered with scabs and boils, who feels that whatever he does is a pain to others and a disgrace to himself. How familiar it all is: the bragging, defiant, aggressive pose to hide one's terrible vulnerability; the lack of self-assurance, the dependence on others, the desperate need for approval, leading to an embarrassing mixture of servility and arrogance; the pathetic eagerness for play, for an escape from the loneliness which he carries with him like a portable cage; the vicious circle of accusations and self-accusations; the exaggerated standards applied to one's own moral conduct which turns life into a long series of Falls into the ninefold inferno of guilt.

Kepler belonged to the race of bleeders, the victims of emotional hemophilia, to whom every injury means multiplied danger, and who nevertheless must go on exposing himself to stabs and slashes. But one customary feature is conspicuously absent from his writings: the soothing drug of self-pity, which makes the sufferer spiritually impotent, and prevents his suffering from bearing fruit. He was a Job who shamed his Lord by making trees grow from his boils. In other words, he had that mysterious knack of finding original outlets for inner pressure; of transforming his torments into creative achievement, as a turbine extracts electric current out of the turbulent stream. His bad eyesight seems the most perfidious trick that fate could inflict on a stargazer; but how is one to decide whether an inborn affliction will paralyze or galvanize? The myopic child, who sometimes saw the world doubled or quadrupled, became the founder of modern optics (the word "diopters" on the oculist's prescription is derived from the title of one of Kepler's books); the man who could see clearly only at a short distance invented the modern astronomical telescope. We shall have occasion to watch the working of this magic dynamo, which transforms pain into achievement and curses into blessings.

4. APPOINTMENT

He was graduated from the Faculty of Arts at the University of Tuebingen at the age of twenty. Then, continuing on the road of his chosen vocation, he matriculated at the Theological Faculty. He studied there for nearly four years, but before he could pass his final examinations, fate intervened. The candidate of divinity was unexpectedly offered the post of a teacher of mathematics and astronomy in Gratz, capital of the Austrian province of Styria.

Styria was a country ruled by a Catholic Hapsburg prince and its predominantly Protestant Estates. Gratz accordingly had both a Catholic university and a Protestant school. When, in 1593, the mathematicus of the latter died, the Governors asked, as they often did, the Protestant University of Tuebingen to recommend a candidate. The Tuebingen Senate recommended Kepler. Perhaps they wanted to get rid of the querulous young man, who had professed Calvinist views and defended Copernicus in a public disputation. He would make a bad priest but a good teacher of mathematics.

33

Kepler was taken by surprise and at first inclined to refuse—"not because I was afraid of the great distance of the place (a fear which I condemn in others), but because of the unexpected and lowly nature of the position, and my scant knowledge in this branch of philosophy."[10] He had never thought of becoming an astronomer. His early interest in Copernicus had been one among many others; it had been aroused, not by an interest in astronomy proper, but by the mystical implications of the sun-centered universe.

Nevertheless, after some hesitation he accepted the offer—mainly, it seems, because it meant financial independence, and because of his inborn love of adventure. He made it a condition, however, that he should be allowed to resume his study of divinity at a later date— but he never did.

The new teacher of astronomy and "Mathematicus of the Province"—a title that went with it—arrived in Gratz in April 1594, at the age of twenty-three. A year later he hit on the idea which would dominate the rest of his life, and out of which his revolutionary discoveries were born.

I have so far concentrated on the emotional life of his childhood and adolescence. I must now briefly speak of his intellectual development. Here again, we have his self-portrait to guide us.

"This man was born destined to spend much time on difficult tasks from which others shrunk. As a boy he precociously attempted the science of versifying. He tried to write comedies and chose the longest poems to learn by heart. . . . His efforts were at first devoted to acrostics and anagrams. Later on he set about various most difficult forms of lyric poetry, wrote a Pindaric lay, dithyrambic poems, and compositions on unusual subjects, such

34

as the resting place of the sun, the sources of rivers, the sight of Atlantis through the clouds. He was fond of riddles and subtle witticisms and made much play with allegories which he worked out to the most minute detail, dragging in farfetched comparisons. He liked to compose paradoxes and . . . loved mathematics above all other studies.

In philosophy he read the texts of Aristotle in the original. . . . In theology he started at once on predestination and fell in with the Lutheran view of the absence of free will. . . . But later on he opposed it. . . . Inspired by his view of divine mercy, he did not believe that any nation was destined to damnation. . . . He explored various fields of mathematics as if he were the first man to do so [and made a number of discoveries], which later on he found to have already been discovered. He argued with men of every profession for the profit of his mind. He jealously preserved all his writings and kept any book he could lay hands on with the idea that they might be useful at some time in the future. He was the equal of Crusius* in his attention to detail, far inferior to Crusius in industry, but his superior in judgment. Crusius collected facts, he analyzed them; Crusius was a hoe, he a wedge. . . ."

In his horoscope he further reports that during his first year at the university he wrote essays on "the heavens, the spirits, the Genii, the elements, the nature of fire, the tides, the shape of the continents, and other things of the same kind."

The last remark about his student days reads:

"At Tuebingen I often defended the opinions of Copernicus in the disputation of the candidates,

* One of Kepler's teachers.

and I composed a careful disputation on the first motion, which consists in the rotation of the earth; then I was adding to this the motion of the earth around the sun for physical, or if you prefer, metaphysical reasons.

If there are living creatures on the moon (a matter about which I took pleasure in speculating after the manner of Pythagoras and Plutarch in a disputation written in Tuebingen in 1593), it is to be assumed that they should be adapted to the character of their particular country."

None of this points as yet in any definite direction. Indeed, his main complaint against himself, which he repeats over and over again, is his "inconsistency, thoughtlessness, lack of discipline, and rashness"; his "lack of persistence in his undertakings, caused by the quickness of his spirit"; his "beginning many new tasks before the previous one is finished"; his "sudden enthusiasms which do not last, for, however industrious he may be, nevertheless he is a bitter hater of work"; his "failure to finish things he has begun."

Again we see that magic dynamo of the psyche at work. The streak of irresponsibility and restlessness in the blood, which turned his father, brother, and uncles into vagabonds who could never settle down in any place or profession, drove Kepler into his unorthodox, often crankish intellectual enterprises, made him into the most reckless and erratic spiritual adventurer of the scientific revolution.

The lectures of this new teacher must have been quite an experience. He thought himself a poor pedagogue because, as he explains in his self-analysis, whenever he got excited—which was most of the time—he "burst into speech without having time to weigh whether he was saying the right thing." His "enthusiasm and eagerness is harmful, and an obstacle to him," because it continually leads him into digressions, because he always thinks of "new words and new subjects, new ways of expressing or proving his point, or even of altering the plan of his lecture or holding back what he intended to say." The fault, he explains, lies in his peculiar kind of memory, which makes him promptly forget everything he is not interested in, but which is quite wonderful in relating one idea to another. "This is the cause of the many parentheses in his lectures when everything occurs to him at once and, because of the turmoil of all these images of thought in his memory, he must pour them out in his speech. On these grounds his lectures are tiring, or at any rate perplexing and not very intelligible."

No wonder that in his first year he had only a handful of students in his class, and in his second, none at all. Barely twelve months after his arrival in Gratz he wrote to his old teacher of astronomy in Tuebingen, Michael Maestlin, that he could not hope to last for another year, and he implored Maestlin to get him a job back home. He felt unhappy, an exile from his sophisticated alma mater among the provincial Styrians. On his arrival, he had been promptly attacked by "Hungarian fever." Besides, religious tension was growing in the town, and made prospects even gloomier.

However, the directors of the school took a more optimistic view. In their report on the new teacher[11] they explained that the absence of students should not be blamed on him, "because the study of mathematics is not every man's affair." They made him give some additional lectures on Virgil and rhetoric, "so that he should not be paid for nothing—until the public is prepared to profit from his mathematics too." The remarkable thing about their reports is their unmitigated approval, not only of Kepler's intellect, but also of his character. He had ". . . given such account of himself that we cannot judge otherwise but that he is, in spite of his youth, a learned and . . . a modest and

. . . a fitting magister and professor." This praise contradicts Kepler's own statement that the head of the school was his "dangerous enemy," because "I did not respect him sufficiently as my superior and disregarded his orders."[12] But young Kepler was as hypochondriacal about his relations to others as he was about his health.

5. ASTROLOGY

Another onerous duty, which he secretly enjoyed, during his four years in Gratz was the publication of an annual calendar of astrological forecasts. This was a traditional obligation imposed on the official mathematicus in Styria and brought an additional remuneration of twenty florins per calendar—which Kepler direly needed at his miserable salary of a hundred and fifty florins (less than seventy-five dollars) per annum.

With his first calendar, Kepler was decidedly lucky. He had prophesied, among other things, a cold spell and an invasion by the Turks. Six months later he reported smugly to Michael Maestlin:

> "By the way, so far the calendar's predictions are proving correct. There is an unheard-of cold in our land. In the Alpine farms people die of the cold. It is reliably reported that when they arrive home and blow their noses, the noses fall off. . . . As for the Turks, on January 1 they devastated the whole country from Vienna to Neustadt, setting everything on fire and carrying off men and plunder."[13]

The successful prophecies of the first calendar contributed more to the popularity of the new mathematicus than his enthusiastic and garbled lectures before an empty classroom. As always in times of crisis, belief in astrology was again on the increase in the sixteenth century, not only among the ignorant, but among eminent scholars. It played an important, and at times a dominant part in Kepler's life. His attitude to it was typical of the contradictions in his character, and of an age of transition.

He started his career with the publication of astrological calendars, and ended it as court astrologer to the Duke of Wallenstein. He did it for a living, with his tongue in his cheek, called astrology "the stepdaughter of astronomy," popular prophecies "a dreadful superstition" and "a sortilegus monkey play."[14] In a typical outburst he wrote: "A mind accustomed to mathematical deduction, when confronted with the faulty foundations [of astrology] resists a long, long time, like an obstinate mule, until compelled by beating and curses to put its foot into that dirty puddle."[15]

But while he despised these crude practices, and despised himself for having to resort to them, he at the same time believed in the possibility of a new and true astrology as an exact empirical science. He wrote a number of serious treatises on astrology as he would understand it, and the subject constantly intrudes even in his classic scientific works. One of these treatises carries, as a motto, "a warning to certain Theologians, Physicians, and Philosophers . . . that, while justly rejecting the stargazers' superstitions, they should not throw out the child with the bath water."[16] For "nothing exists nor happens in the visible sky that is not sensed in some hidden manner by the faculties of Earth and Nature: [so that] these faculties of the spirit here on earth are as much affected as the sky itself."[17] And again: "That the sky does something to man is obvious enough; but what it does specifically remains hidden."[18] In other words, Kepler regarded the current astrological practices as quackery, but only to the extent to which a modern physician distrusts an unproven slimming diet, without doubting for a moment the influence of diet on health and figure. "The belief in the effect of the constellations derives in the first place from experience, which is so

39

convincing that it can be denied only by people who have not examined it."[19]

We have seen that in his self-analysis, in spite of its astoundingly modern introspective passages and acute characterizations of his family, all main events and character attributes were derived from the planetary constellations. But, on reflection, what other explanation was there available at the time? To a questing mind without an inkling of the processes by which heredity and environment shape a man's character, astrology, in one form or another, was the obvious means of relating the individual to the universal whole, by making him reflect the all-embracing constellation of the world, by establishing an intimate sympathy and correspondence between microcosmos and macrocosmos: "The natural soul of man is not larger in size than a single point, and on this point the form and character of the entire sky is potentially engraved, even if it were a hundred times larger."[20] Unless predestination alone were to account for everything, making further inquiry into the Book of Nature pointless, it was only logical to assume that man's condition and fate were determined by the same celestial motions which determined the weather and the seasons, the quality of the harvest, the fertility of animal and plant. In a word, astrological determinism, to a scientific mind like Kepler's, was the forerunner of biological and psychological determinism.

Even as a child he was fascinated by the problem of why he had become what he had become. We remember the passage in his self-analysis: "In theology I started at once on predestination and fell in with the Lutheran view of the absence of free will." But he quickly repudiated it. When he was thirteen, "I wrote to Tuebingen asking that a certain theological treatise be sent to me, and one of my comrades upbraided me thus: 'Bachelor, does't thou too suffer from doubts about predestination?' "[21] The mystery of "why am I what I am?"

must have been experienced with particular intensity by a precocious and unhappy adolescent in that century of awakening, when individual consciousness was emerging from the collective consciousness of the medieval bee-hive hierarchy, where queens and warriors, workers and drones, had all inhabited their ordained cubbyholes in existence. But if there was no predestination, how was one to explain the differences in character and personality, talent and worth, between members of the same race, all descended from Adam; or between young Johannes himself, the infant prodigy, and his epileptic brother? Modern man has an explanation of sorts in terms of chromosomes and genes, adaptive responses and traumatic experiences; sixteenth-century man could only search for an explanation in the state of the universe as a whole at the moment of his conception or birth, as expressed by the constellation of earth, planets, and stars.

The difficulty was to find out how exactly this influence worked. That "the sky does something to man" was self-evident; but specifically what? "Truly in all my knowledge of astrology I know not enough with certainty that I should dare to predict with confidence any specific thing."[21a] Yet he never gave up hope.

"No man should hold it to be incredible / that out of the astrologers' foolishness and blasphemies / some useful and sacred knowledge may come / that out of the unclean slime / may come a little snail / or mussel / or oyster or eel, all useful nourishments; / that out of a big heap of lowly worms / may come a silkworm / and lastly / that in the evil-smelling dung / a busy hen may find a decent corn / nay, a pearl or a golden corn / if she but searches and scratches long enough."[22]

There is hardly a page in Kepler's writings—some

twenty solid volumes in folio—that is not alive and kicking.

And, gradually, a vision did indeed emerge out of the confusion. At twenty-four, he wrote to a correspondent:

"In what manner does the countenance of the sky at the moment of a man's birth determine his character? It acts on the person during his life in the manner of the loops which a peasant ties at random around the pumpkins in his field: they do not cause the pumpkin to grow, but they determine its shape. The same applies to the sky: it does not endow man with his habits, history, happiness, children, riches, or a wife, but it molds his condition. . . ."[23]

Thus only the pattern is cosmically determined, not any particular event; within that pattern, man is free. In his later years, this *Gestalt* concept of cosmic destiny became more abstract and purified from dross. The individual soul, which bears the potential imprint of the entire sky, reacts to the light coming from the planets according to the angles they form with each other, and the geometrical harmonies or disharmonies that result—just as the ear reacts to the mathematical harmonies of music, and the eye to the harmonies of color. This capacity of the soul to act as a cosmic resonator has a mystic and a causal aspect: on the one hand it affirms the soul's affinity with the anima mundi; on the other, it makes it subject to strictly mathematical laws. At this point, Kepler's particular brand of astrology merges into his all-embracing and unifying Pythagorean vision of the harmony of the spheres.

chapter two

The "Cosmic Mystery"

1. THE PERFECT SOLIDS

From the frustrations of his first year in Gratz, Kepler escaped into the cosmological speculations he had playfully pursued in his Tuebingen days. But now these speculations were becoming both more intense, and more mathematical in character. A year after his arrival—more precisely, on July 9, 1595, for he has carefully recorded the date—he was drawing a figure on the blackboard for his class, when an idea suddenly struck him with such force that he felt he was holding the key to the secret of creation in his hand. "The delight that I took in my discovery," he wrote later, "I shall never be able to describe in words."[1] It determined the course of his life, and remained his main inspiration throughout it.

The idea, centuries old in its essence, was that the universe is built around certain symmetrical figures—triangle, square, pentagon, etc.—which form its invisible skeleton, as it were. Before going into detail, it will be better to explain at once that the idea itself was completely false; yet it led eventually to Kepler's Laws, the demolition of the antique universe on wheels, and the birth of modern cosmology. The pseudo-discovery which started it all is expounded in Kepler's first book, the

*Mysterium Cosmographicum,** which he published at
the age of twenty-five.

In the preface to the work, Kepler explained how he
came to make his "discovery." While still a student in
Tuebingen, he had heard from his teacher in astronomy,
Maestlin, about Copernicus, and agreed that the sun
must be in the center of the universe—"for physical, or
if you prefer, for metaphysical reasons." He then began
to wonder why there existed just six planets "instead of
twenty or a hundred," and why the distances and velocities of the planets were what they were. Thus started
his quest for the laws of planetary motion.

At first he tried to find whether one orbit might per-
chance be twice, three or four times as large as another.
"I lost much time on this task, on this play with numbers; but I could find no order either in the numerical
proportions or in the deviations from such proportions."
He warns the reader that the tale of his various futile
efforts "will anxiously rock thee hither and thither like
the waves of the sea." Since he got nowhere, he tried
"a startlingly bold solution": he inserted an auxiliary
planet between Mercury and Venus, and another between Jupiter and Mars, both supposedly too small to
be seen, hoping that now he would get some sensible
sequence of ratios. But this did not work either; nor did
various other devices which he tried.

"I lost almost the whole of the summer with this
heavy work. Finally I came close to the true facts on
a quite unimportant occasion. I believe Divine
Providence arranged matters in such a way that

* The full title reads: A Forerunner (Prodromus) to Cos-
mographical Treatises, containing the Cosmic Mystery of the
admirable proportions between the Heavenly Orbits and the
true and proper reasons for their Numbers, Magnitudes, and
Periodic Motions, by Johannes Kepler, Mathematicus of the
Illustrious Estates of Styria, Tuebingen, anno 1596.

what I could not obtain with all my efforts was given to me through chance; I believe all the more that this is so as I have always prayed to God that he should make my plan succeed, if what Copernicus had said was the truth."[2]

The occasion of this decisive event was the aforementioned lecture to his class, in which he had drawn, for quite different purposes, a geometrical figure on the blackboard. The figure showed (I must describe it in a simplified manner) a triangle fitted between two circles;

FIG. 1.

in other words, the outer circle was circumscribed around the triangle, the inner circle inscribed into it.

As he looked at the two circles, it suddenly struck him that their ratios were the same as those of the orbits of Saturn and Jupiter. The rest of the inspiration came in a flash. Saturn and Jupiter are the "first" (i.e., the two outermost) planets, and "the triangle is the first figure in geometry. Immediately I tried to inscribe into the next interval between Jupiter and Mars a square, between Mars and Earth a pentagon, between Earth and Venus a hexagon. . . ."

It did not work—not yet, but he felt that he was quite

close to the secret. "And now I pressed forward again. Why look for two-dimensional forms to fit orbits in space? One has to look for three-dimensional forms— and, behold, dear reader, now you have my discovery in your hands!"

The point is this. One can construct any number of regular polygons in a two-dimensional plane; but one can construct only a limited number of regular solids in three-dimensional space. These "perfect solids," of which all faces are identical, are: (1) the tetrahedron (pyramid) bounded by four equilateral triangles; (2) the cube; (3) the octahedron (eight equilateral triangles); (4) the dodecahedron (twelve pentagons); and (5) the icosahedron (twenty equilateral triangles).

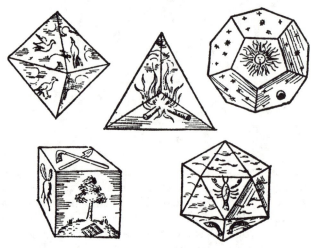

FIG. 2. *The five regular solids.*
From Harmonices Mundi, *Liber II* (1619).

They were also called the "Pythagorean" or "Platonic" solids, because early Greek thinkers ascribed mystical properties to them. Being perfectly symmetrical, each can be *inscribed* into a sphere, so that all its vertices (corners) lie on the surface of the sphere. Similarly, each can be *circumscribed* around a sphere, so that the sphere touches every face in its center. It is an odd fact, inherent in the nature of three-dimensional space, that (as Euclid proved) the number of regular solids is limited to these five forms. Whatever shape you choose as a face, no other perfectly symmetrical solid can be constructed except these five. Other combinations just cannot be fitted together.

So there existed only five perfect solids—and five intervals between the planets! It was impossible to believe that this should be by chance, and not by divine arrangement. It provided the complete answer to the question why there were just six planets "and not twenty or a hundred." And it also answered the question why the distances between the orbits were as they were. They had to be spaced in such a manner that the five solids could be exactly fitted into the intervals, as an invisible skeleton or frame. And, lo, they fitted! Or at least they seemed to fit, more or less. Into the orbit, or sphere, of Saturn he inscribed a cube; and into the cube another sphere, which was that of Jupiter. Inscribed in that was the tetrahedron, and inscribed in it the sphere of Mars. Between the spheres of Mars and Earth came the dodecahedron; between Earth and Venus the icosahedron; between Venus and Mercury the octahedron. Eureka! The mystery of the universe was solved by young Kepler, teacher at the Protestant school in Gratz. (Plate I.)

"It is amazing!" Kepler informs his readers. "Although I had as yet no clear idea of the order in which the perfect solids had to be arranged, I nevertheless succeeded . . . in arranging them so happily that later on, when I checked the matter over, I had nothing to alter. Now I no longer regretted the lost time; I no longer tired of my work; I shied from no computation, however difficult. Day and

47

night I spent with calculations to see whether the proposition that I had formulated fitted the Copernican orbits or whether my joy would be carried away by the winds. . . . Within a few days everything fell into its place. I saw one symmetrical solid after the other fit in so precisely between the appropriate orbits that if a peasant were to ask you on what kind of hook the heavens are fastened so that they don't fall down, it will be easy for thee to answer him. Farewell!"[3]

We had the privilege of witnessing one of the rare recorded instances of a false inspiration, a supreme hoax of the Socratic *daimon*, the inner voice that speaks with such infallible, intuitive certainty to the deluded mind. That unforgettable moment before the figure on the blackboard carried the same inner conviction as Archimedes' "Eureka" or Newton's flash of insight about the falling apple. But there are few instances where a delusion led to momentous and true scientific discoveries and yielded new laws of nature. This is the ultimate fascination of Kepler—both as an individual and as a case history. For Kepler's misguided belief in the five perfect bodies was not a passing fancy, but remained with him, in a modified version, to the end of his life, showing all the symptoms of a paranoid delusion; and yet it functioned as the *vigor motrix*, the spur of his immortal achievements. He wrote the *Mysterium Cosmographicum* when he was twenty-five, but he published a second edition of it a quarter-century later, toward the end, when he had done his lifework, discovered his three Laws, destroyed the Ptolemaic universe, and laid the foundations of modern cosmology. The dedication to this second edition, written at the age of fifty, betrays the persistence of the *idée fixe*.

"Nearly twenty-five years have passed since I published the present little book. . . . Although I

movement. The overture consists of the *Introduction to the Reader*, which I have already discussed, and the first chapter, which is an enthusiastic and lucid profession of faith in Copernicus.[4] It was the first unequivocal, public commitment by a professional astronomer which appeared in print fifty years after Copernicus' death, and the beginning of his posthumous triumph.[5] Galileo, Kepler's senior by six years, and astronomers like Maestlin were still either silent on Copernicus, or agreed with him only in cautious privacy. Kepler had intended to add to his chapter a proof that there was no contradiction between Holy Scripture and the Copernican doctrine that the earth moved around the sun; but the head of the theological faculty in Tuebingen, whose official consent to the publication of the book had to be obtained, directed him to leave out any theological reflections and—in the tradition of the famous Osiander preface—to treat the Copernican hypothesis as a purely formal, mathematical one.* Kepler accordingly postponed his theological apologia to a later work, but otherwise did the exact opposite of what he was advised to do, by proclaiming the Copernican system to be literally, physically, and incontrovertibly true, "an inexhaustible treasure of truly divine insight into the wonderful order of the world and all bodies therein." It sounded like a fanfare in praise of the brave new heliocentric world. The arguments in its favor that Kepler adduced could mostly be found in Rheticus' *Narratio Prima*, which Kepler reprinted as an appendix to the *Mysterium*, to save his readers the labor of toiling through Copernicus' unreadable book.

After this overture, Kepler gets down to his "principal proof": that the planetary spheres are separated from each other, or fenced in, as it were, by the five perfect

* It was Kepler himself who, a few years later, discovered that the preface to Copernicus' book, *De Revolutionibus*, was written by a clergyman named Osiander and not by Copernicus.

solids. (He does not mean, of course, that the solids are really present in space, nor does he believe in the existence of the spheres themselves, as we shall see.) The "proof" consists, roughly, in the deduction that God could create only a perfect world, and since only five symmetrical solids exist, they are obviously meant to be placed between the six planetary orbits "where they fit in perfectly." In fact, however, they do not fit at all, as he was soon to discover, to his woe. Also, there are not six planets but nine (not to mention the small fry of asteroids between Jupiter and Mars), but at least Kepler was spared in his lifetime the discovery of the three others, Uranus, Neptune, and Pluto.

In the next six chapters (III to VIII), it is explained to us why there are three planets outside and two inside the earth's orbit; why that orbit is placed just where it is; why the cube lies between the two outermost planets and the octahedron between the two innermost; what affinities and sympathies exist between the various planets and the various solids, and so on—all this by a priori deductions derived straight from the Creator's secret thoughts, and supported by reasons so fantastic that one can hardly believe one is listening to one of the founders of modern science. Thus, for instance, "the regular solids of the first order [i.e., those which lie outside the earth's orbit] have it in their nature to stand upright, those of the second order to float. For, if the latter are made to stand on one of their sides, the former on one of their corners, then in both cases the eye shies from the ugliness of such a sight." By this kind of argument young Kepler succeeds in proving everything that he believes and in believing everything that he proves. The ninth chapter deals with astrology, the tenth with numerology, the eleventh with the geometrical symbolism of the zodiac; in the twelfth, he alludes to the Pythagorean harmony of the spheres, searching for correlations between his perfect solids and the harmonic intervals in music—but it is merely one more arabesque to the dream. On this note ends the first half of the book.

The second half is different. I have talked of a work in two movements, because they are written in different moods and keys, and are held together only by their common leitmotif. The first is medieval, aprioristic, and mystical; the second, modern and empirical. The *Mysterium* is the perfect symbol of the great watershed.

The opening paragraph of the second half must have come as a shock to his readers.

"What we have so far said served merely to support our thesis by arguments of probability. Now we shall proceed to the astronomical determination of the orbits and to geometrical considerations. If these do not confirm the thesis, then all our previous efforts have doubtless been in vain."[6]

So all the divine inspiration and a priori certitude were merely "probabilities"; and their truth or falsehood was to be decided by the observed facts. Without transition, in a single startling jump, we have traversed the frontier between metaphysical speculation and empirical science.

Now Kepler gets down to brass tacks: the checking of the proportions of his model of the universe against the observed data. Since the planets do not revolve around the sun in circles, but in oval-shaped orbits (which Kepler's First Law, years later, identified as ellipses), each planet's distance from the sun varies within certain limits. This variation (or eccentricity) he accounted for by allotting to each planet a spherical shell of sufficient thickness to accommodate the oval orbit between its walls (see model on Plate I). The inner wall represents the planet's minimum distance from the sun, the outer wall its maximum distance. The spheres, as already mentioned, are not considered as physically real, but merely as the limits of space allotted to each orbit.

The thickness of each shell and the intervals between them were laid down in Copernicus' figures. Were they spaced in such a way that the five solids could be exactly fitted between them? In the preface, Kepler had confidently announced that they could. Now he found that they could not. There was fairly good agreement for the orbits of Mars, Earth, and Venus, but not for Jupiter and Mercury. The trouble with Jupiter, Kepler dismissed with the disarming remark that "nobody will wonder at it, considering the great distance." As for Mercury, he frankly resorted to cheating.[7] It was a kind of Wonderland croquet through mobile celestial hoops.

In the following chapters Kepler tried various methods to explain away the remaining discrepancies. The fault must lie either in his model or in the Copernican data; and Kepler naturally preferred to blame the latter. First, he discovered that Copernicus had placed into the center of the world not really the sun, but the center of the earth's orbit, "in order to save himself trouble and so as not to confuse his diligent readers by dissenting too strongly from Ptolemy."[8] Kepler undertook to remedy this, hoping thereby to obtain more favorable *Lebensraum* (living space) for his five solids. His mathematical knowledge was as yet insufficient for this task, so he turned for help to his old teacher, Maestlin, who willingly complied. The new figures did not help Kepler at all; yet he had at one stroke, and almost inadvertently, shifted the center of the solar system where it belonged. It was the first momentous by-product of the phantom chase.

His next attempt to remedy the disagreement between his dream and the observed facts concerned the moon. Should her orbit be included into the thickness of the earth's sphere or not? He explained frankly to his dear readers that he would choose the hypothesis which best fits his plan; he will tuck the moon into the earth's shell, or banish her into the outer darkness, or let her orbit stick halfway out, for there are no a priori reasons in favor of either solution. (Kepler's a priori proofs were mostly found a posteriori.) But fiddling with the moon did not help either, so young Kepler proceeded to a frontal attack against the Copernican data. With admirable impertinence he declared them to be so unreliable that Kepler's own figures would be strongly suspect if they agreed with Copernicus'. Not only were the tables unreliable; not only was Copernicus inexact in his observations, as reported by Rheticus (from whom Kepler quotes long, damning passages); but Copernicus also cheated.

"How human Copernicus himself was in adopting figures which within certain limits accorded with his wishes and served his purpose; this the diligent reader of Copernicus may test by himself. . . . He selects observations from Ptolemy, Walter, and others with a view to making his computations easier, and he does not scruple to neglect or to alter occasional hours in observed time and quarter-degrees of angle."[9]

Twenty-five years later, Kepler himself amusedly commented on his first challenge of Copernicus:

"After all, one approves of a toddler of three who decides that he will fight a giant."[10]

So far, in the first twenty chapters of his book, Kepler had been concerned with finding reasons for the number and spatial distribution of the planets. Having satisfied himself (if not his readers) that the five solids provided all the answers, and that existing discrepancies were due to Copernicus' faulty figures, he now turned to a different and more promising problem, which no astronomer before him had raised. He began to look for a mathematical relation between a planet's distance from the sun, and the length of its "year"—that is, the time it needed for a complete revolution.

These periods were, of course, known since antiquity with considerable precision. In round figures, Mercury needs three months to complete a revolution, Venus seven and a half months, the earth a year, Mars two years, Jupiter twelve years, and Saturn thirty years. Thus, the greater the planet's distance from the sun, the longer it takes to complete a revolution, but this is only roughly true: an exact mathematical ratio was lacking. Saturn, for instance, is twice as far out in space as Jupiter, and should therefore take twice as long to complete a circuit—that is, twenty-four years; but Saturn in fact takes thirty. The same is true of the other planets. As we travel from the sun outward into space, the motion of the planets along their orbits gets slower and slower. (To make the point quite clear: they not only have a longer way to travel to complete a circuit, but they also travel at a slower rate along it. If they traveled at the same rate, Saturn, with a circuit twice as long as Jupiter's, would take twice as long to complete it; but it takes two and a half times as long.)

Nobody before Kepler had asked the question *why* this should be so, as nobody before him had asked why there are just six planets. As it happens, the latter question proved scientifically sterile,* the former immensely fertile. Kepler's answer was, that there must be *a force emanating from the sun* which drives the planets round their orbits. The outer planets move more slowly because this driving force diminishes in ratio to distance, "as does the force of light."

It would be difficult to overestimate the revolutionary significance of this proposal. For the first time since antiquity, an attempt was made, not only to *describe*

---

* At least, our mathematical tools are as yet inadequate for tackling the genesis and morphology of the solar system. Much depends on asking the right question at the right time.

heavenly motions in geometrical terms, but to assign them a *physical cause*. We have arrived at the point where astronomy and physics meet again, after a divorce which lasted for two thousand years. This reunion of the two halves of the split mind produced explosive results. It led to Kepler's three Laws, the pillars on which Newton built the modern universe.

Again we are in the fortunate position of being able to watch, as in a slow-motion film, how Kepler was led to taking that decisive step. In the key passage from the *Mysterium Cosmographicum* which follows, the index numbers are Kepler's own, and refer to his notes in the second edition.

"If we want to get closer to the truth and establish some correspondence in the proportions [between the distances and velocities of the planets] then we must choose between these two assumptions: either the souls[ii] which move the planets are the less active the farther the planet is removed from the sun, or there exists only one moving soul[iii] in the center of all the orbits, that is the sun, which drives the planet the more vigorously the closer the planet is, but whose force is quasi-exhausted when acting on the outer planets because of the long distance and the weakening of the force which it entails."[11]

To this passage Kepler made, in the second edition, the following notes:

"(ii). That such souls do not exist I have proved in my *Astronomia Nova.*

"(iii). If we substitute for the word 'soul' the word 'force' then we get just the principle which underlies my physics of the skies in the *Astronomia Nova.* . . . For once I firmly believed that the motive force of a planet was a soul. . . . Yet as I re-

56

flected that this cause of motion diminishes in proportion to distance, just as the light of the sun diminishes in proportion to distance from the sun, I came to the conclusion that this force must be something substantial—'substantial' not in the literal sense but . . . in the same manner as we say that light is something substantial, meaning by this an unsubstantial entity emanating from a substantial body."[12]

We are witnessing the hesitant emergence of the modern concepts of "forces" and "radiating energies" which are both material and non-material, and, generally speaking, as ambiguous and bewildering as the mystical concepts which they have come to replace. As we watch the working of the mind of Kepler (or Paracelsus, Gilbert, Descartes) we are made to realize the fallacy of the belief that at some point between the Renaissance and the Enlightenment, man shook off the "superstitions of medieval religion" like a puppy getting out of the water, and started on the bright new road of Science. Inside these minds, we find no abrupt break with the past, but a gradual transformation of the symbols of their cosmic experience—from *anima motrix* into *vis motrix*, moving spirit into moving force, mythological imagery into mathematical hieroglyphics—a transformation which never was and, one hopes, never will be entirely completed.

The details of Kepler's theory were again all wrong. The driving force which he attributed to the sun has no resemblance to gravity; it is rather like a whip which lashes the sluggish planets along their paths. As a result, Kepler's first attempt to formulate the law relating planetary distances with periods was so obviously wrong that he had to admit it.[18] He added wistfully:

57

"Though I could have foreseen this from the beginning, I nevertheless did not want to withhold
from the reader this spur to further efforts. Oh, that
we could live to see the day when both sets of figures agree with each other! . . . My only purpose
was that others may feel stimulated to search for
that solution toward which I have opened the
path."[14]

But it was Kepler himself who found the correct solution, toward the end of his life: it is his Third Law. In
the second edition of the *Mysterium*, he added a note
to the phrase, "Oh, that we could live to see the
day. . . ." It reads:

"We have lived to see this day after twenty-two
years and rejoiced in it, at least I did; I trust that
Maestlin and many other men . . . will share in my
joy."[15]

The closing chapter of the *Mysterium* is a return to
the medieval shore of the Keplerian torrent of thought.
It is described as "the dessert after this substantial
meal," and concerns the constellations of the sky on the
first and last days of the world. We are given a fairly
promising horoscope for the Creation—which started on
Sunday, April 27, 4977 B.C.; but about the last day Kepler says modestly, "I did not find it possible to deduce
an end of the motions from inherent reasons."
On this childish note ends Kepler's first book, the
dream of five perfect solids determining the scheme of
the universe. The history of thought knows many barren
truths and fertile errors. Kepler's error turned out to be
of immense fertility. "The direction of my whole life,
of my studies and works, has been determined by this
one little book," he wrote a quarter-century later.[16]
"For nearly all the books on astronomy which I have
published since then were related to one or the other

of the main chapters in this little book and are more thorough expositions or completions of it."[17] Yet he also had an inkling of the paradoxical nature of all this, for he added:

"The roads by which men arrive at their insights into celestial matters seem to me almost as worthy of wonder as those matters in themselves."[18]

3. BACK TO PYTHAGORAS

One crucial question was left unexplained in the previous chapters. What exactly was it that so forcefully attracted Kepler, when he was still a student of theology, to the Copernican universe? In his self-analysis he expressly stated that it was not interest in astronomy proper, that he was converted "by physical, or if you prefer, metaphysical reasons"; and he repeats this statement almost verbatim in the preface to the *Mysterium*. These "physical or metaphysical reasons" he explains differently in different passages; but the gist of them is, that the sun must be in the center of the world because "he" is the symbol of God the Father, the source of light and heat, the generator of the force which drives the planets in their orbits, and because a sun-centered universe is geometrically simpler and more satisfactory. These seem to be four different reasons, but they form a single, indivisible complex in Kepler's mind, a new Pythagorean synthesis of mysticism and science.

We remember that to the Pythagoreans and Plato the animating force of the Deity radiated from the center of the world outward, until Aristotle banished the First Mover to the periphery of the universe. In the Copernican system, the sun again occupied the place of the Pythagorean Central Fire, but God remained outside, and the sun had neither divine attributes nor any physical influence on the motions of the planets. In Kepler's universe, all mystic attributes and physical powers are

59

centralized in the sun, and the First Mover is returned to the focal position where he belongs. The visible universe is the symbol and "signature" of the Holy Trinity: the sun represents the Father; the sphere of the fixed stars, the Son; the invisible forces which, emanating from the Father, act through interstellar space, represent the Holy Ghost.

"The sun in the middle of the moving stars, himself at rest and yet the source of motion, carries the image of God the Father and Creator. . . . He distributes his motive force through a medium which contains the moving bodies even as the Father creates through the Holy Ghost."[19]

The fact that space has three dimensions is itself a reflection, a "signature" of the mystic Trinity:

"And thus are bodily things, thus are *materia corporea* represented in *tertia quantatis specie trium dimensionum*."[20]

The unifying truth between the mind of God and the mind of man is represented for Kepler, as it was for the Pythagorean brotherhood, by the eternal and ultimate truths of "divine geometry."

"Why waste words? Geometry existed before the Creation, is co-eternal with the mind of God, *is God himself* (what exists in God that is not God himself?); geometry provided God with a model for the Creation and was implanted into man, together with God's own likeness—and not merely conveyed to his mind through the eyes."[21]

But if God created the world after a geometrical model, and endowed man with an understanding of geometry, then it must be perfectly feasible, young Kepler thought, to deduce the whole blueprint of the universe by pure a priori reasoning, by reading the mind of the Creator, as it were. The astronomers are "the priests of God, called to interpret the Book of Nature," and surely priests have a right to know the answers.

If Kepler's evolution had stopped here, he would have remained a crank. But I have already pointed out the contrast between the a priori deductions in the first part of the book and the modern scientific approach of the second. This coexistence of the mystical and the empirical, of wild flights of thought and dogged, painstaking research, remained, as we shall see, the main characteristic of Kepler from his early youth to his old age. Other men living on the watershed displayed the same dualism, but in Kepler it was more pointed and paradoxical, carried to extremes verging on insanity. It accounts for the incredible mixture in his works of recklessness and pedantic caution, his irritability and patience, his naïveté and philosophical depth. It emboldened him to ask questions which nobody had dared to ask without trembling at their audacity, or blushing at their apparent foolishness. Some of them appear meaningless to the modern mind. The others led to the reconciliation of earth physics with sky geometry, and were the beginning of modern cosmology. That some of his own answers were wrong does not matter. As in the case of the Ionian philosophers of the heroic age, the philosophers of the Renaissance were perhaps more remarkable for the revolutionary nature of the questions they asked than for the answers they proposed. Paracelsus and Bruno, Gilbert and Tycho, Kepler and Galileo formulated some answers which are still valid; but first and foremost they were giant question masters. *Post factum*, however, it is always difficult to appreciate the originality and imagination it required to ask a question which had not been asked before. In this respect, too, Kepler holds the record.

Some of his questions were inspired by a medieval brand of mysticism, and yet proved to be amazingly

fertile. The shifting of the First Mover from the periphery of the universe into the physical body of the sun,
symbol of the Godhead, prepared the way for the concept of a gravitational force, symbol of the Holy Ghost,
which controls the planets. Thus a purely mystical inspiration was the root out of which the first rational
theory of the dynamics of the universe developed, based
on the secular trinity of Kepler's Laws.

Equally astonishing was the fertility of Kepler's
errors—starting with a universe built around the five
solids, and ending with a universe governed by musical
harmonies. This process, of error begetting truth, is il-
luminated by Kepler's own comments on the *Mysterium
Cosmographicum*. Written twenty-five years later, they
are contained in his notes to the second edition, to which
I have repeatedly referred. In complete contrast to his
claim that the book was written as if under the dicta-
tion of a "heavenly oracle," and represented "an obvious
act of God," Kepler's notes castigate its errors with acid
sarcasm. The book starts, as we remember, with an "Outline of my Principal Proof" and Kepler's comment starts
with, "Woe to me, here I blundered." The ninth chapter deals with the "sympathies" between the five solids
and the individual planets; in the notes it is dismissed
as a mere "astrological fancy." Chapter 10, "On the
Origin of Privileged Numbers," is described in the notes
as "empty chatter"; Chapter 11, "Concerning the Positions of the Regular Solids and the Origin of the
Zodiac," is qualified in the notes as "irrelevant, false,
and based on illegitimate assumptions." On Chapter 17,
concerning the orbit of Mercury, Kepler's comments
are: "this is not at all true," "the reasoning of the whole
chapter is wrong." The important twentieth chapter,
"On the Relation between the Motions and Orbits," in
which the Third Law is foreshadowed, is dismissed as
faulty "because I used uncertain ambiguous words instead of arithmetical method." The twenty-first chapter,

which discusses discrepancies between theory and observation, is attacked in the notes in an almost unfairly
petulant manner; e.g.: "This question is superfluous.
. . . Since there is no discrepancy, why did I have to
invent one?"

Yet the notes to this chapter contain two remarks in
a different key.

> "If my false figures came near to the facts, this
> happened merely by chance. . . . These comments
> are not worth printing. Yet it gives me pleasure to
> remember how many detours I had to make, along
> how many walls I had to grope in the darkness of
> my ignorance until I found the door which lets in
> the light of truth. . . . In such manner did I dream
> of the truth."[22]

By the time he had finished with his notes to the
second edition (which amount to approximately the
same length as the original work) the old Kepler had
demolished practically every point in the book of the
young Kepler—except its subjective value to him as the
starting point of his long journey, a vision which, though
faulty in every detail, was "a dream of truth": "inspired
by a friendly God." The book indeed contained the
dreams, or germs, of most of his later discoveries—as by-
products of its erroneous central idea. But in later years,
as the notes show, this *idée fixe* was intellectually neu-
tralized by so many qualifications and reserves that it
could do no harm to the working of his mind; while his
irrational belief in its basic truth remained, emotion-
ally, the motive power behind his achievements. The
harnessing to a rational pursuit of the immense psychic
energies derived from an irrational obsession seems to
be another secret of genius, at least of genius of a certain type. It may also explain the distorted view of their
own achievements so frequently found among them.
Thus in Kepler's notes to the *Mysterium* he proudly re-

fers to some minor discoveries in his later works, but there is not one single mention of the first and second of his immortal Laws, which every schoolboy associates with his name. The notes are chiefly concerned with the planetary orbits, yet the fact that these are ellipses (Kepler's First Law) is nowhere mentioned; it was as if Einstein, in his old age, had been discussing his work without mentioning relativity. Kepler set out to prove that the solar system was built like a perfect crystal around the five divine solids, and discovered, to his chagrin, that it was dominated by lopsided and undistinguished curves; hence his unconscious taboo about the word "ellipse," his blind spot for his greatest achievement, and his clinging to the shadow of the *idée fixe*.[23] He was too sane to ignore reality, but too mad to value it.

A modern scholar remarked about the scientific revolution, "One of the most curious and exasperating features of this whole magnificent movement is that none of its great representatives appears to have known with satisfying clarity just what he was doing or how he was doing it."[24] Kepler, too, discovered his America, believing that it was India.

But the urge that drove him on was not aimed at any practical benefit. In the labyrinth of Kepler's mind, the thread of Ariadne is his Pythagorean mysticism, his religious-scientific quest for a harmonious universe governed by perfect crystal shapes or perfect chords. It was this thread that led him, through abrupt turns and dizzy gyrations, in and out of dead ends, to the first exact laws of nature, to the healing of the millennial rift between astronomy and physics, to the mathematization of science. Kepler said his prayers in the language of mathematics, and distilled his mystic faith into a mathematician's Song of Songs.

"Thus God himself / was too kind to remain idle / and began to play the game of signatures / signing his likeness unto the world: therefore I chance to think / that all nature and the graceful sky are / symbolized in the art of Geometria. . . . / Now, as God the maker play'd / he taught the game to Nature / whom he created in his image: / taught her the selfsame game / which he played to her. . . ."[25]

Here at last was the jubilant refutation of Plato's cave. The living world no longer is a dim shadow of reality, but Nature's dance to which God sets the tune. Man's glory lies in his understanding of the harmony and rhythm of the dance, an understanding made possible through his divine gift of thinking in numbers.

". . . these figures pleased me because they are quantities, that is, something which existed before the skies. For quantities were created at the beginning, together with substance; but the sky was only created on the second day. . . .[26] The ideas of quantities have been and are in God from eternity, they are God himself; they are therefore also present as archetypes in all minds created in God's likeness. On this point both the pagan philosophers and the teachers of the Church agree."[26a]

By the time Kepler wrote down this credo, the first stage in the young pilgrim's progress was completed. His religious doubts and anxieties had been transformed into the mystic's mature innocence—the Holy Trinity into a universal symbol, his craving for the gift of prophecy into the search for ultimate causes. The sufferings of a mange-eaten, chaotic childhood had left a sober thirst for universal law and harmony; memories of a brutal father may have influenced his vision of an abstract God, without human features, bound by mathematical rules which admitted of no arbitrary acts.

His physical appearance had undergone an equally

radical change: the adolescent with the bloated face and spindly limbs had grown into a sparse, dark, wiry figure, charged with nervous energy, with chiseled features and a somewhat Mephistophelian profile, belied by the melancholia of the soft, shortsighted eyes. The restless student who had never been able to finish what he began had changed into a scholar with a prodigious capacity for work, for physical and mental endurance, and a fanatical patience unequaled in the annals of science.

chapter three

Growing Pains

1. THE COSMIC CUP

The inspiration about the five perfect solids had come to Kepler when he was twenty-four, in July 1595. During the next six months he had worked feverishly on the *Mysterium*. He reported on every stage of his progress to Maestlin in Tuebingen, pouring out his ideas in long letters and asking for his former teacher's help, which Maestlin gave in a grumbling but generous manner.

Michael Maestlin was twenty years Kepler's senior, yet was to outlive him. A contemporary engraving shows him as a bearded worthy with a jovial and somewhat vacant face. He had held the chair of mathematics and astronomy at Heidelberg, then at his native Tuebingen, and was a competent teacher with a solid academic reputation. He had published a textbook on astronomy of the conventional type, based on the Ptolemaic system, although in his lectures he spoke with admiration of Copernicus, and thus ignited the spark in young Kepler's inflammable mind. After the manner of good-natured mediocrities who know and accept their own limitations, he had a naïve admiration for the genius of his former pupil and went to considerable trouble to help him, though with an occasional growl at Kepler's unceasing demands. When the book was finished and the Senate of Tuebingen asked for Maestlin's expert opinion, he enthusiastically recommended that it should be published; and when permission was granted, he supervised the printing himself. This, in those days, was practically a full-time job; as a result Maestlin was reprimanded by the university Senate for neglecting his own work. He complained about this to Kepler in understandably peeved tones; Kepler replied, among his usual effusions of gratitude, that Maestlin shouldn't worry about the reprimand since, by seeing the *Mysterium* through print, Maestlin had acquired immortal fame.

By February 1596 the rough draft of the book was completed and Kepler asked his superiors in Gratz for leave of absence to visit his native Wuerttemberg and make arrangements for its publication. He asked for two months, but stayed away for seven, as he had become involved in a typically Keplerian chimera. He had persuaded Frederick, Duke of Wuerttemberg, to have a model of the universe, incorporating the five perfect solids, made in the shape of a drinking cup. "A childish or fatal craving for the favor of princes," as he later confessed, had driven him to Stuttgart, to Frederick's court, to whom he explained his idea in a letter.

"Since the Almighty granted me last summer a major *inventum* in astronomy, after lengthy, unsparing toil and diligence; which same *inventum* I have explained in a special booklet which I am willing to publish any time; which whole work and demonstration thereof can be fittingly and gracefully represented by a drinking cup of an ell in diameter which then would be a true and genuine likeness of the world and model of the creation insofar as human reason may fathom, and the like of which has never before been seen or heard of by any man; therefore I have postponed the preparation of such a model or its showing to any man to the present time of my arrival from Styria, intending to put this true and correct model of the world before the eyes of your Grace, as my natural sovereign, for him to see it as the first man on earth."[1]

Kepler went on to suggest that the various parts of the cup should be made by different silversmiths, and then fitted together, to make sure that the cosmic secret would not leak out. The signs of the planets could be cut in precious stones—Saturn in diamond, Jupiter in jacinth, the moon a pearl, and so on. The cup would serve seven different kinds of beverage, conducted by concealed pipes from each planetary sphere to seven taps on its rim. The sun would provide a delicious aqua vita; Mercury, brandy; Venus, mead; the moon, water; Mars, a strong vermouth; Jupiter, "a delicious new white wine"; and Saturn "a bad old wine or beer," "whereby those ignorant in astronomical matters could be exposed to shame and ridicule." Assuring Frederick that in ordering the cup he would do a favor to the arts and a service to God Almighty, Kepler remained Frederick's obedient servant, hoping for the best.

The duke wrote on the margin of Kepler's letter, "Let him first make a model of copper and when we see it and decide that it is worth being made in silver, the means shall not want." Kepler's letter was dated February 17, and the duke's answer was transmitted to him on the next day; Frederick's imagination had obviously caught on. But Kepler had no money to make a copper model, as he resentfully conveyed to the duke in his next letter; instead, he settled down to the Herculean task of making a paper model of all the planetary orbits and the five perfect solids in between. He labored day and night for a week; years later he nostalgically remarked that it had been quite a pretty model, made out in paper of different colors, with all the orbits in blue.

When the paper monster was finished, he sent it to the duke, apologizing for its clumsiness and huge dimensions. Again promptly on the next day, the duke ordered his chancellery to ask for the expert opinion of Professor Maestlin. The good Maestlin wrote to Frederick that Kepler's cup would represent a "glorious work of erudition," and the duke wrote on the margin: "Since this is so, we are content that the work should be executed."

But apparently it had been easier for God to build the world around the five polyhedra than for the silversmiths to execute a copy of it. Besides, Frederick did not want the cosmic mystery in the form of a drinking cup, but to have it encased in a celestial globe. Kepler made another paper model, left it with the silversmith, and in September returned to Gratz, having wasted nearly six months at Frederick's court. But the duke would not drop the project, and it dragged on for several years. In January 1598 Kepler wrote to poor Maestlin (who now served as the go-between), "If the duke agrees, it would be best to break up the whole junk and refund the silver to him. . . . The thing is hardly worth while. . . . I started it too ambitiously."[2] But six months later, he submitted via Maestlin a new project. The cup, which had turned into a globe, was now to turn into a mobile planetarium, driven by a clockwork. The description of it occupied ten printed pages in folio. Kepler informed the duke that a Frankfurt mathematician, Jacob Cuno, had offered to construct a planetarium which would reproduce the heavenly motions "within an error of one degree for the next six or ten thousand years"; but, Kepler explained, such a machine would be too large and costly, and proposed a more modest one, guaranteed for a century only. "For it is not to be hoped (apart from the Last Judgment) that such a work would remain undisturbed in one place over a hundred years. Too many wars, fires, and other changes are wont to occur."[3]

The correspondence went on for another two years; then the subject was at last mercifully forgotten. But this quixotic escapade inevitably reminds one of the ill-fated vagabondages of his father, uncle, and brother. He worked off his innate restlessness in bold imaginations and plain drudgery; but from time to time some residual poison in his blood would make him break out in a rash and momentarily turn the sage into a clown. This fact is painfully evident in the tragicomedy of Kepler's first marriage.

2. MARRIAGE

Before his journey to Wuerttemberg, Kepler's friends in Gratz had found a prospective bride for the young mathematicus, in the daughter of a rich mill owner, twice widowed at the age of twenty-three. Barbara Muehleck had been married at sixteen, against her wish, to a middle-aged cabinetmaker who had died after two years; then to an elderly, widowed pay clerk who brought into the marriage a bunch of misshapen children, chronic illness, and, after his timely demise, was found to have defrauded money in his trust. Barbara, described by Kepler as "simple of mind and fat of body," now lived with her parents, who could not have very high expectations of her future. Yet when Kepler presented his suit through two respectable middlemen (a school inspector and a deacon) the proud miller refused, on the grounds that he could not entrust Barbara and her dowry to a man of such lowly standing and miserable pay. This was the beginning of long and sordid negotiations conducted by Kepler's friends with the family.

When he left for Stuttgart nothing was settled, but in the spring his friends wrote to him that his suit had been accepted, advised him to hurry home, and to bring with him from Ulm "some truly good silk cloth, or at least of the best double taffeta, sufficient for complete robes for thyself and the bride." But Kepler, too busy with his cosmic silver cup, delayed his return, and by the time he got back to Gratz, Frau Barbara's father had changed his mind again. Kepler seems not to have been unduly perturbed, but the indefatigable friends continued their efforts; the dean of the school and even the Church authorities joined in—"and so they vied with one another to assault the minds, now of the widow, now of her father, took them by storm and arranged for me a new date for the nuptials. Thus, with one blow, all my plans for beginning another life collapsed."[4]

The marriage. took place on April 27, 1597, "under a calamitous sky," as the horoscope indicated. He was somewhat comforted by the arrival of the first printed copies of the *Mysterium Cosmographicum*, but not even that event was all joy; he had to buy two hundred copies of the book for cash to compensate the printer for the risk; and the author's name in the catalogue of the Frankfurt Book Fair was transformed, by misprint, from Keplerus into Repleus.

Kepler's attitude to marriage in general, and to his own wife in particular, is expressed in several letters with shocking frankness. The first is addressed to Maestlin and dated two weeks before the wedding. It occupies nearly six pages in folio, of which only the last speaks of the impending great event.

"I ask you only one favor, that you should be close to me in your prayers on my wedding day. My financial situation is such that should I die within the next year hardly anybody could leave a worse situation after him. I am obliged to spend a big sum of my own, for it is the custom here to celebrate marriages splendidly. If, however, God prolongs my life, I shall be bound and constricted to this place. . . . For my bride possesses here estates, friends, and a prosperous father; it seems that after a few years I would not need my salary any longer. . . . Thus I shall be unable to leave this province except if a public or a private misfortune intervened. A public misfortune it would be if the country ceased to be safe for Lutherans, or if the Turks, who have already massed six hundred thousand men, invaded it. A private misfortune it would be if my wife died."[5]

Not a word is said about the person of his betrothed or his feelings for her. But in another letter, written two years later, he blames her horoscope for her "rather sad and unlucky fate. . . . In all dealings she is confused and inhibited. Also she gives birth with difficulty. Everything else is in the same vein."[6]

After her death, he described her in even more depressing terms. She had known how to make a favorable impression on strangers, but at home she had been different. She resented her husband's lowly position as a stargazer and understood nothing of his work. She read nothing, not even stories, only her prayer book, which she devoured day and night. She had "a stupid, sulking, lonely, melancholy complexion." She was always ailing and weighed down with melancholia. When his salary was withheld, she refused to let him touch her dowry, even to pawn a cup or to put her hand into her private purse.

"And because, due to her constant illness, she was deprived of her memory, I made her angry with my reminders and admonitions, for she would have no master and yet often was unable to cope herself. Often I was even more helpless than she, but in my ignorance persisted in the quarrel. In short, she was of an angry nature, and uttered all her wishes in an angry voice; this incited me to provoke her, I regret it, for my studies sometimes made me thoughtless; but I learned my lesson, I learned to have patience with her. When I saw that she took my words to heart, I would rather have bitten my own finger than to give her further offense. . . ."[7]

Her avarice made her neglect her appearance; but she lavished everything on the children because she was a woman "entirely imprisoned by maternal love"; as for her husband—"not much love came my way." She nagged not only him but also the servant wenches, and "could never keep a wench." When he was working she would interrupt him to discuss her household problems. "I may have been impatient when she failed to understand and went on asking me questions, but I never called her a fool, though it may have been her understanding that I considered her a fool, for she was very touchy."[8]

Nine months after the wedding their first child was born, a little boy. After two months the child died of cerebral meningitis, and the next, a little girl, died after a month, of the same disease. Frau Barbara bore three more children, of which one boy and one girl survived.

Altogether, their marriage lasted fourteen years; Barbara died at the age of thirty-seven, with a distraught mind. The marriage horoscope had shown a *coelo calamitoso*, and in predicting disaster Kepler's horoscopes were nearly always right.

3. LIMBERING UP

When, in the spring of 1597, the *Mysterium* at last appeared in print, the proud young author sent copies to all leading scholars he could think of, including Galileo and Tycho de Brahe. There existed as yet no scientific journals (nor, happy days, book reviewers); on the other hand, there was an intensive exchange of letters among scholars and a luxuriant international academic grapevine. By these means the unknown young man's book created a certain stir; though not the earthquake which its author expected, yet remarkable enough if we consider that the average number of scientific (and pseudo-scientific) books published in Germany in a single year was well over a thousand.[9]

Growing Pains

But the response was not surprising. Astronomy, from Ptolemy to Kepler, had been a purely descriptive geography of the sky. Its task was to provide maps of the fixed stars, timetables of the motions of sun, moon, and planets, and of such special events as eclipses, oppositions, conjunctions, solstices, equinoxes, and the rest. The physical causes of the motions, the forces of nature behind them, were not the astronomer's concern. Whenever necessary, a few epicycles (circles whose centers move on circles) were added to the existing machinery of wheels—which did not matter much, since they were fictional anyway, and nobody believed in their physical reality. The hierarchy of cherubim and seraphim who were supposed to keep the wheels turning was, since the end of the Middle Ages, regarded as another polite, poetic fiction. Thus the physics of the sky had become a complete blank. There were events but no causes, motions but no moving forces. The astronomer's task was to observe, describe, and predict, not to search for causes—"theirs not to reason why." Aristotelian physics, which made any rational and causal approach to the heavenly phenomena unthinkable, was on the wane, but it had left only a vacuum behind it. Ears were still ringing with the vanished song of the star-spinning angels, but all was silence. In that fertile silence the unformed, stammering voice of the young theologian-turned-astronomer obtained an immediate hearing.

Opinions were divided according to the philosophy of the scholars. The modern and empirically minded, such as Galileo in Padua and Praetorius in Altdorf, rejected Kepler's mystical a priori speculations and with them the whole book, without realizing the explosive new ideas hidden among the chaff. Galileo, especially, seems to have been prejudiced from the beginning against Kepler, of which we shall hear more later on.

Those, however, who lived on the other side of the

75

watershed, who believed in the ageless dream of an a priori deduction of the cosmic order, were enthusiastic and delighted. Most of all, of course, the endearing Maestlin, who wrote to the Tuebingen Senate:

"The subject is new and has never before occurred to anybody. It is most ingenious and deserves in the highest degree to be made known to the world of learning. Who has ever dared before to think, and much less to try to expose and explain a priori and, so to speak, out of the hidden knowledge of the Creator, the number, order, magnitude, and motion of the spheres? But Kepler has undertaken and successfully done just this. . . . Henceforth [astronomers] shall be freed from the necessity of exploring the dimensions of the spheres a posteriori, that is, by the method of observations (many of which are inexact and not to say doubtful) after the manner of Ptolemy and Copernicus, because now the dimensions have been established a priori. . . . Whereby the computation of the movements will become much more successful. . . ."[10]

In a similar vein enthused Limneus in Jena, who congratulated Kepler, all students of astronomy, and the whole learned world that "at last the old and venerable [Platonic] method of philosophy had been resurrected."[11]

In a word, the book which contained the seeds of the new cosmology was welcomed by the "reactionaries," who did not see its implications, and rejected by the "moderns," who did not see them either. Only one man took a middle course and, while rejecting Kepler's wild speculations, immediately realized his genius: the most outstanding astronomer of the day, Tycho de Brahe.

But Kepler had to wait for three years until he met Tycho, became his assistant, and started on his true life-

76

work. During these three years (1597–99) he at last got down to a serious study of mathematics, of which he had still been shockingly ignorant when he wrote the *Mysterium*, and he undertook a motley variety of scientific and pseudo-scientific researches. It was a kind of limbering up before the great contest.

The first task he set himself was to find direct confirmation of the earth's motion around the sun by proving the existence of stellar parallax—that is, a shift in the apparent position of the fixed stars according to the earth's position on its annual journey. He pestered, in vain, all his correspondents to help him with observations, and at last decided to take a peep for himself; but his "observatory" consisted of a simple staff suspended on a rope from the ceiling: "it comes from a workshop like the huts of our forebears—hold your laughter, friends, who are admitted to this spectacle."[12] Even so, it would have been sufficiently precise to show the variation of half a degree, which Kepler expected, in the positions of the polar star as seen from extreme points of the earth's path. But there was no variation; the starry sky remained immutable, poker-faced. This meant either that the earth stood still, or that the size of the universe (that is, the radius of the sphere of fixed stars) was much larger than previously assumed. To be precise, its radius must be at least five hundred times the distance of the earth from the sun. This works out at 2400 million miles, a trifle by our, but not a lot even by Kepler's, standards; only about five times more than he had expected.[13] However, even if much better instruments failed to show a parallax, indicating that the stars are quite inconceivably distant, in the eyes of God the universe would still have a reasonable size; only man's physical stature would shrink. But this would not diminish his moral stature, "otherwise the crocodile or the elephant would be nearer to His heart than man, because they are larger. With the help of this and similar intellectual pills, we shall perhaps be able to digest this monstrous bite."[14] In fact, no pill has been discovered since to digest the lump of infinity.

Other problems which occupied him were his first researches into optics, out of which eventually a new science was to emerge; investigations of the moon's orbit, of magnetism, of meteorology—he started a weather diary which he kept up for twenty or thirty years; of Old Testament chronology, and the like. But dominating all these interests was his search for a mathematical law of the harmony of the spheres—a further development of his *idée fixe*.

In the *Mysterium*, Kepler had tried to build his universe around the five Pythagorean solids. Since the theory did not quite fit the facts, he now tried to build it around the musical harmonies of the Pythagorean scale. The combination of these two ideas led, twenty years later, to his great work *Harmonice Mundi* (*Harmony of the World*), which contains the third of Kepler's Laws; but the groundwork for it was laid during his last years in Gratz.

The moment this new idea had occurred to him, his letters resounded with jubilant "Eurekas": "Fill the skies with air, and they will produce true and real music." But as he began to compute the details of his cosmic musical box, he ran into increasing difficulties. He was never short of an excuse for ascribing to any pair of planets the musical interval which approximately happened to fit it; when things became sticky, he asked the shade of Pythagoras for help—"unless the soul of Pythagoras has migrated into mine." He managed to construct a system of sorts, but its inadequacies were obvious to himself. The principal trouble was that a planet does not move at uniform speed, but faster when it is close to the sun, slower when away from it. Accordingly it does not "hum" on a steady pitch, but alternates between a lower and a higher note. The interval between the two notes depends on the lopsidedness or "eccentricity" of the planet's orbit. But the eccentricities were only inaccurately known. It was the same difficulty he had come up against when he had tried to define the thickness of the spherical shells between his perfect solids, which also depended on the eccentricities. How could you build a series of crystals, or a musical instrument, without knowing the measurements? There was only one man alive in the world who possessed the exact data which Kepler needed: Tycho de Brahe.

All his hopes became focused now on Tycho, and Tycho's observatory at Uraniburg, the new wonder of the world.

"Let all keep silence and hark to Tycho, who has devoted thirty-five years to his observations. . . . For Tycho alone do I wait; he shall explain to me the order and arrangement of the orbits. . . . Then I hope I shall one day, if God keeps me alive, erect a wonderful edifice."[15]

Thus he knew that the building of that edifice still lay in the distant future, though in his euphoric moments he claimed to have it already completed. During his manic periods, the discrepancies between theory and fact appeared to him as contemptible details, which could be smoothed over by a little cheating; yet the other half of his divided self humbly acknowledged the duty of pedantic accuracy and patient observation. With one eye he was reading the thoughts of God; the other squinted enviously at Tycho's shining armillary spheres.

But Tycho refused to publish his observations until he had completed his own theory. He jealously guarded his treasure, volumes of figures, the result of a lifetime of work.

"Any single instrument of his," young Kepler wrote bitterly, "cost more than my and my whole family's fortune put together. . . . My opinion of Tycho is this: he is superlatively rich, but he knows not how to make proper use of it, as is the case with most rich people. Therefore, one must try to wrest his riches from him."[16]

In this outcry, Kepler had revealed his intentions toward Tycho de Brahe a year before they met for the first time.

4. WAITING FOR TYCHO

Had Kepler not succeeded in getting hold of Tycho's treasure, he could never have discovered his planetary laws. Now, Newton was born only twelve years after Kepler's death, and he could not have arrived at his synthesis without the planetary laws. They could only be discovered with Tycho's help; and by the time Kepler met him, Tycho had only eighteen months left to live. If it was Divine Providence which timed their meeting, it chose a rather perverse method: Kepler was hounded out of Gratz and into the arms of Tycho by religious persecution. Though he always endeavored to read the thoughts of God, he never offered his thanks for this Machiavellian stratagem.

That last year in Gratz—the last of the century—was indeed not easy to endure. The young Archduke Ferdinand of Hapsburg (later Emperor Ferdinand II) was determined to cleanse the Austrian provinces of the Lutheran heresy. In the summer of 1598, Kepler's school was closed down, and in September all Lutheran preachers and schoolmasters were ordered to leave the province within eight days or forfeit their lives. Only one among them received permission to return, and that was Kepler. His exile, the first, lasted less than a month.

The reasons why an exception was made with him are rather interesting. He himself says[17] that the archduke

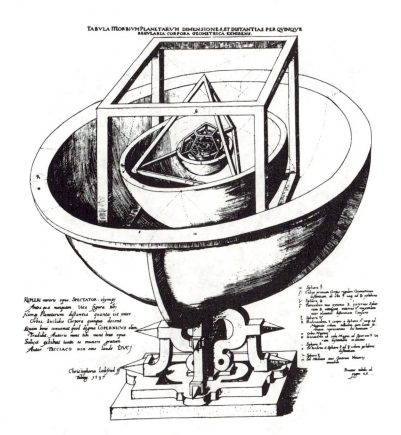

PLATE I. Model of the universe;
the outermost sphere is Saturn's. From
Mysterium Cosmographicum (1597, edition of 1621).

PLATE II.   Johannes Kepler.
Redrawn from steel engraving by A. Weger.

PLATE III. This is a fine and accurate reconstruction
of Tycho Brahe's observatory as painted by
Heinrich Hansen in 1882; the original is in the
Art Gallery of Frederiksborg Castle. Denmark.

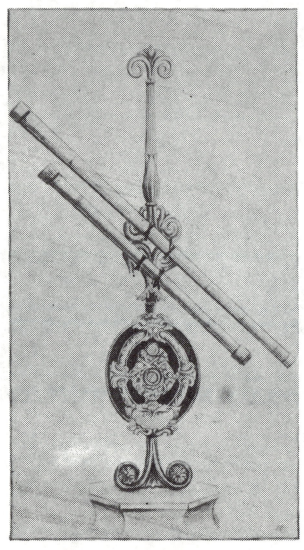

PLATE IV. Two of Galileo's telescopes
as displayed in the Museum of History of Sciences
in Florence. In the center of the medallion
is a cracked lens.

was "pleased with my discoveries" and that this was the reason for his favor at his court; besides, as a mathematicus he occupied a "neutral position" which set him apart from the other teachers. But it was not as simple as that. Kepler had a powerful ally behind the scenes: the Jesuit order.

Two years previously the Catholic chancellor of Bavaria, Herwart von Hohenburg, amateur philosopher and patron of the arts, had asked Kepler, among other astronomers, for his opinion on certain chronological problems. It was the beginning of a lifelong correspondence and friendship between the two men. Herwart tactfully advertised his protective interest in the Protestant mathematicus by sending his letters to Kepler via the Bavarian envoy at the Emperor's court in Prague, who forwarded them to a Capuchin father at Ferdinand's court in Gratz; and he instructed Kepler to use the same channels. In his first letter to Herwart,[18] Kepler wrote delightedly, "Your letter so impressed some men in our government that nothing more favorable to my reputation could have happened."

It was all done with great subtlety; yet on later occasions Catholic and especially Jesuit influences were more openly active on behalf of Kepler's welfare. There seem to have been three reasons for this benevolent cabal. First, a scholar was still to some extent regarded as a sacred cow amidst the turmoil of religious controversy. Second, the Jesuits, following in the steps of the Dominicans and Franciscans, were beginning to play a leading part in science and especially in astronomy—quite apart from the fact that it enabled their missionaries in distant countries to make a great impression by predicting eclipses and other celestial events. And, last, Kepler himself disagreed with certain points of Lutheran doctrine, which made his Catholic friends hope—though in vain—that he might become a convert. He did a certain amount of fence straddling, but he refused to change sides, even when he was excommunicated by his own church, as we shall hear; and when he suspected that Herwart was counting on his conversion, Kepler wrote to him:

"I am a Christian, the Lutheran creed was taught me by my parents, I took it unto myself with repeated searchings of its foundations, with daily questionings, and I hold fast to it. Hypocrisy I have never learned. I am in earnest about Faith and I do not play with it."[19]

It was the outburst of a man of basic integrity, forced to swim in the troubled waters of his time. He was as sincere in matters of religion as circumstances permitted him to be; at any rate, his deviations from the straight path were perhaps not greater than those of his orbits from God's five perfect solids.

Kepler was, then, made an exception and permitted to return from exile in October 1599. Since his school had been closed down, he could devote most of his time to his speculations on the harmony of the spheres; yet he knew that the reprieve was only a temporary one, and that his days in Gratz were numbered. He sank into a profound depression, deepened by the death of his second child; in a despairing letter he asked Maestlin, in August 1599, for his help in finding a job at home in Protestant Wuerttemberg.

"The hour could not have been more propitious; but God has offered this fruit, too, only to take it away again. The child died of a cerebral meningitis (exactly as its brother a year ago) after thirty-five days. . . . If its father should follow soon, his fate would not be unexpected. For everywhere in Hungary bloody crosses have appeared on the bodies of men and similar bloody signs on the gates of houses, on benches and walls, which history shows to be a sign of a general pestilence. I am, as far as I know, the first person in our town to see a small cross on my left foot, the color of which passes from bloody red to yellow. The spot is on the foot, where the back of the foot curves into the instep, halfway between the toes and the end of the shinbone. I believe it is just the spot where the nail was hammered into the foot of Christ. Some carry, I am told, marks in the shape of drops of blood in the hollow of the hand. But so far this form has not appeared on me. . . .

The ravages of dysentry kill people of all ages here, but particularly children. The trees stand with dry leaves on their crowns as if a scorching wind had passed over them. Yet it was not the heat that so disfigured them, but worms. . . ."[20]

He had the worst fears. There was talk of torture for heretics, even of burnings. He was fined ten dalers for burying his child according to Lutheran rites: "half of it was remitted at my request, but the other half I had to pay before I was allowed to carry my little daughter to her grave." If Maestlin cannot get him a job at once, would he at least let him know about the present cost of living in Wuerttemberg: "How much wine costs and how much wheat, and how things stand regarding the supply of delicatessen (for my wife is not in the habit of living on beans)."

But Maestlin knew that his university would never give a job to the unruly Kepler, and he was getting thoroughly fed up with Kepler's unceasing demands and badgerings; the more so, as Kepler had followed up his SOS with the foolish remark:

"Of course, nobody would expel me; the most intelligent among the members of the Diet are the most fond of me, and my conversation at meals is much sought after."[21]

No wonder that Maestlin underestimated the urgency of the situation, and delayed for five months before he answered with an evasive and grumpy epistle. "If only you had sought the advice of men wiser and more experienced in politics than I, who am, I confess, as unexperienced in such matters as a child."[22]

Only one hope remained: Tycho. The previous year, Tycho, in a letter, had expressed the hope that Kepler would "someday" visit him. Though Kepler was panting and pining for "Tycho's treasure," the invitation was couched in too general terms, and the journey too long and costly. Now, however, it was no longer a matter of scientific curiosity for Kepler, but of the urgent necessity of finding a new home and a livelihood.

Tycho, in the meantime, had been appointed imperial mathematicus by Rudolph II, and had taken up residence near Prague. Kepler's long-awaited opportunity came when a certain Baron Hoffmann, councilor to the Emperor, had to return from Gratz to Prague, and agreed to take him along in his suite. The date of Kepler's departure for the meeting with Tycho is, by the courtesy of history, easy to remember: it was January 1, anno Domini 1600.

84

chapter four

Tycho de Brahe

1. THE QUEST FOR PRECISION

Johannes Kepler was a pauper who came from a family of misfits; Tycho de Brahe was a *grand seigneur* from the Hamlet country, the scion of truculent and quixotic noblemen of pure Danish stock. His father had been governor of Helsingborg Castle, which faces Elsinore across the Sund; his Uncle Joergen, a country squire and vice-admiral.

This Uncle Joergen, being childless, had extracted a promise from his brother, the governor, that if he had a son, Joergen could adopt him and bring him up as his own. Nature seemed to sanction this arrangement, for, in 1546, the Governor's wife bore him twin sons; but unfortunately one of them was stillborn, and the father went back on his promise. Joergen, a true, headstrong Brahe, waited until another son was born to his brother, then kidnaped the firstborn, Tyge—Tycho. The governor, also in true Brahe fashion, threatened murder, but quickly cooled off and generously consented to the *fait accompli*, knowing that the child would be well looked after and inherit some of Joergen's fortune. This came indeed to pass, and sooner than expected, for while Tyge was still a student, his foster-father met an untimely and glorious end. He had just returned from a naval battle against the Swedes, and was riding in the suite of his

85

king over the bridge joining Copenhagen to the royal
castle, when the good king, Ferdinand II, fell into the
water. Joergen, the vice-admiral, jumped after him,
saved his king, and died of pneumonia.

Whether Tyge suffered a traumatic shock by being
kidnaped from his cradle, we cannot know; but the blood
of the Brahes and his education by the irascible vice-admiral must have sufficed to turn him into an eccentric
in the grand style. This was visible at first glance, even
in his physical appearance: for if Tycho was born with
a silver spoon in his mouth, he later acquired a nose of
silver and gold. As a student, he fought a duel with another noble Danish youth, in the course of which part
of Tyge's nose was sliced off. According to a contemporary account,[1] the quarrel originated in a dispute as to
which of the two noble Danes was the better mathematician. The lost piece, which seems to have been the
bridge of the nose, was replaced by a gold and silver
alloy, and Tycho is said to have always carried a kind of
snuffbox "containing some ointment or glutinous com-
position which he frequently rubbed on his nose."[2] In
his portraits, the nose appears as a too-rectilinear, cub-
istic feature among the curves of a large, bald, egg-
shaped head, set between the cold, haughty eyes and the
aggressively twirled handlebar mustache.

True to the family tradition, young Tyge was intended to take up the career of a statesman, and was
accordingly sent at thirteen to study rhetoric and philosophy at the University of Copenhagen. But at the
end of his first year he witnessed an event which made
an overwhelming impression on him and decided the
whole future course of his life. It was a partial eclipse
of the sun, which, of course, had been announced beforehand, and it struck the boy as "something divine that
men could know the motions of the stars so accurately
that they were able a long time beforehand to predict
their places and relative positions."[3] He immediately be-

gan to buy books on astronomy, including the collected works of Ptolemy for the considerable sum of two Joachims-Thaler. From now onward his course was set and he never swerved from it.

Why did that partial eclipse, which was not at all spectacular as a sight, have such a decisive impact on the boy? The great revelation for him, Gassendi tells us, was the *predictability* of astronomical events—in total contrast, one might speculate, to the unpredictability of a child's life among the temperamental Brahes. It is not much of a psychological explanation, but it is worth noting that Brahe's interest in the stars took from the beginning a quite different, in fact almost opposite, direction from both Copernicus' and Kepler's. It was not a speculative interest, but a passion for exact observation. Starting on Ptolemy at fourteen, and making his first observation at seventeen, Tycho took to astronomy at a much earlier age than those two. The timid Canon Copernicus had found a refuge from a life of frustrations in the secret elaboration of his system; Kepler resolved the unbearable miseries of his youth in his mystic harmony of the spheres. Tycho was neither frustrated nor unhappy, only bored and irritated by the futility of a Danish nobleman's existence among, in his own words, "horses, dogs, and luxury"; and he was filled with naïve wonder at the contrasting soundness and reliability of the stargazers' predictions. He took to astronomy not as an escape or metaphysical life belt, but rather as a full-time hobby of an aristocrat in revolt against his milieu. His later life seems to confirm this interpretation, for he entertained kings on his wonder island, but the mistress of the house, with whom he begot a large family of children, was a woman of low caste to whom he was not even married in church.

After three years at Copenhagen, the vice-admiral thought that it was time for Tyge to go to a foreign university, and sent him, accompanied by a tutor, to

Leipzig. The tutor was Anders Soerensen Vedel, who later became famous as the first great Danish historian, translator of Saxo Grammaticus and collector of Nordic sagas. Vedel was then twenty, only four years older than his charge; he had received instructions to cure young Tyge of his unseemly preoccupation with astronomy and lead him back to studies more fitting for a nobleman. Tyge had bought a small celestial globe to learn the names of the constellations, but he had to hide it under his blanket; and when he added to this a cross staff, he could use it only when his tutor was asleep. After a year, Vedel, realizing that Tyge was star-struck beyond remedy, gave in, and the two remained lifelong friends.

After Leipzig, Tycho continued his studies at the universities of Wittenberg, Rostock, Basle, and Augsburg until his twenty-sixth year, all the time collecting, and later designing, bigger and better instruments for observing the planets. Among these was a huge quadrant of brass and oak, thirty-eight feet in diameter and turned by four handles—the first of a series of fabulous instruments which were to become a wonder of the world. Tycho never made any epoch-making discovery except one, which made him the father of modern observational astronomy; but that one discovery has become such a truism to the modern mind that it is difficult to see its importance. The discovery was that astronomy needed *precise* and *continuous* observational data.

Copernicus recorded only twenty-seven observations of his own in the whole *Book of Revolutions;* for the rest he relied on the data of Hipparchus, Ptolemy, and others. This had been the general practice up to Tycho. It was taken for granted that planetary tables must be exact, as far as possible, for calendrical and navigational purposes; but apart from the limited data required for these practical reasons, the necessity for precision was not at all realized. This attitude, which is all but incomprehensible to the modern mind, was partly due to the Aristotelian tradition, with its emphasis on qualities instead of quantitative measurement; within that mental framework only a crank could be interested in precision for precision's sake. Besides, and more specifically, a geometry of the skies consisting of cycles and epicycles did not require a great many, or even very precise, observational data, for the simple reason that a circle is defined when its center and a single point of its circumference are known, or, if the center is unknown, by three points of its circumference alone. Hence it was, by and large, sufficient to determine the positions of a planet at a few characteristic points of its orbit, and then to arrange one's epicycles and deferents in the way most favorable to "save the phenomena." If one projects one's mind back to the other side of the watershed, Tycho's devotion to measurements, to fractions of minutes of arc, appears highly original. No wonder that Kepler called him the Phoenix of Astronomy.

On the other hand, if Tycho was ahead of his time, he was only a step ahead of Kepler. We have seen how Kepler was pining for Tycho's observations, for precise data on mean distances and eccentricities. A century earlier, Kepler would probably have rested on the laurels of his solution of the cosmic mystery without bothering about those small disagreements with observed facts; but this metaphysical cavalier attitude toward facts was on the wane among the advanced minds of the time. Ocean navigation, the increasing precision of magnetic compasses and clocks, and the general progress in technology created a new climate of respect for hard fact and exact measurement. Thus, for instance, the debate between the Copernican and Ptolemaic systems was no longer pursued by theoretical arguments alone; both Kepler and Tycho independently decided to let experiment be the arbiter, and tried to determine by measurement whether a stellar parallax existed or not.

One of the reasons for Tycho's quest for precision was, in fact, his desire to check the validity of the Copernican system. But this was perhaps rather the rationalization of a deeper urge. Meticulous patience, precision for precision's sake, was for him a form of worship. His first great experience had been the awe-stricken realization that astronomic events could be exactly predicted; his second was of the opposite kind. On August 17, 1563, at the age of seventeen, while Vedel was asleep, he noticed that Saturn and Jupiter were so close together as to be almost indistinguishable. He looked up his planetary tables and discovered that the Alphonsine tables were a whole month in error regarding this event, and the Copernican tables by several days. This was an intolerable and shocking state of affairs. If the stargazers, of whose low company his family so disapproved, could not do better, let a Danish nobleman show them how a proper job is done.

And show them he did, with methods and gadgets the like of which the world had never seen.

2. THE NEW STAR

At the age of twenty-six, Tycho considered his education complete, and returned to Denmark. For the next five years, till 1575, he lived first on the family estate at Knudstrup, then with an uncle, Steen Bille, the only one in the family who approved of Tycho's perverse hobby. Steen had founded the first paper mill and glassworks in Denmark, and dabbled a lot in alchemy, in which Tycho assisted him.

Like Kepler, Tycho stood with one foot in the past and was devoted both to alchemy and astrology. Like Kepler, he became a court astrologer and had to waste much of his time with the casting of horoscopes for patrons and friends; like Kepler, he did it with his tongue in his cheek, despised all other astrologers as quacks, and yet was profoundly convinced that the stars influenced man's character and destiny, though nobody quite knew how. Unlike Kepler's, however, his belief in astrology derived not from mysticism—which was completely alien to his domineering nature—but from stark superstition.

The great event of these years, an event that was discussed all over the world and that established, at a single stroke, Tycho's fame as the leading astronomer of his time, was the new star of 1572. In Tycho's life, all the decisive landmarks were sky marks: the eclipse of the sun when he was fourteen, which brought him to astronomy; the conjunction of Jupiter and Saturn when he was seventeen, which made him realize its insufficiencies; the new star when he was twenty-six; and the comet of 1577, five years later. Of all these, the new star was the most important.

On the evening of November 11, 1572, Tycho was walking from Steen's alchemist laboratory back to supper when, glancing at the sky, he saw a star brighter than Venus at her brightest, in a place where no star had been before. The place was a little to the northwest of the familiar "W"—the constellation of Cassiopeia, which then stood near the zenith. The sight was so incredible that he literally did not believe his eyes; he called at first some servants, and then several peasants, to confirm the fact that there really was a star where no star had any business to be. It was there, all right, and so bright that later on, people with sharp eyes could see it even in the middle of the day. And it remained in the same spot for eighteen months.

Other astronomers besides Tycho had seen the new star in the first days of November. It was then in full blaze; in December it began very slowly to fade, but it ceased to be visible only by the end of March, the year after that. The world had never seen or heard the like since the year 125 B.C., when Hipparchus, according to the second book of Pliny's *Natural History*, had seen a new star appear in the sky.

The sensational importance of the event lay in the fact that it contradicted the basic doctrine—Aristotelian, Platonic, and Christian—that all change, all generation and decay, was confined to the immediate vicinity of the earth, the sublunary sphere; whereas the distant eighth sphere in which the fixed stars were located was immutable from the day of Creation to eternity. The only known exception in history was the appearance of the new star of Hipparchus; but that had been very long ago, and one could explain it away by assuming that Hipparchus had merely seen a comet (which was then considered an atmospheric phenomenon in the sublunary region).

Now, what distinguishes a fixed star from a planet, or a comet, or a meteor, is the fact that it is "fixed": apart from its participation in the daily rotation of the firmament as a whole, it does not move. As soon as that bright new cuckoo egg appeared on the tip of the celestial "W," far outshining the legitimate stars in its nest, stargazers all over Europe feverishly tried to determine whether it moved or not. If it did, it was not a real star and academic science was saved; if it did not, the world had to be thought afresh.

Maestlin in Tuebingen who, though one of the leading astronomers of the time, seems to have possessed no instruments whatsoever, held a thread at arm's length from his eyes in such a way that it passed through the new star and two other fixed stars. When, after a few hours, the three were still in the same straight line, he concluded that the new star did not move.[4] Thomas Digges in England used a similar method, and came to the same result; others found a displacement, but only a small one, due, of course, to the errors of their coarse instruments. This was Tycho's great opportunity, and he fully rose to it. He had just finished a new instrument—a sextant with arms five and a half feet long, joined by a bronze hinge, with a metallic arc scale graduated to single minutes and, as a novelty, a table of figures designed to correct the errors of the instrument. It was like a heavy gun compared to the slings and catapults of his colleagues. The result of Tycho's observations was unequivocal: the new star stood still in the sky.

All Europe was agog, both with the cosmological and astrological significance of the event. The new star had appeared only about three months after the massacre of French Protestants on St. Bartholomew's Night; no wonder that in the flood of pamphlets and treatises on the star, it was mostly regarded as a sinister omen. The German painter George Busch, for instance, explained that it was really a comet condensed from the rising vapors of human sins, which had been set afire by the wrath of God. It created a kind of poisonous dust (rather like the fallout from a hydrogen bomb) which was drifting down on people's heads and caused all sorts of evil, such as "bad weather, pestilence, and Frenchmen." The more serious astronomers, with few exceptions, tried to explain the star away from the eighth sky by calling it a tailless comet, ascribing to it a slow motion, and using other subterfuges which made Tycho contemptuously talk of *O coecos coeli spectatores*—O blind watchers of the sky.

The next year, he published his first book: *De Stella Nova*. He hesitated some time before publishing it, because he had not yet quite overcome the idea that the writing of books was an undignified occupation for a nobleman. The book is a hodgepodge of tedious prefatory letters, calendrical and meteorological diaries, astrological predictions, and versified outpourings, including an eight-page "Elegy to Urania"; but it contained in twenty-seven pages an exact description of Tycho's observations of the new star, and of the instrument with which the observations were made—twenty-seven pages of "hard, obstinate facts," which alone would suffice to establish his lasting fame.

Five years later he gave Aristotelian cosmology the *coup de grâce*, by proving that the great comet of 1577 was also not a sublunary phenomenon, as comets had previously been regarded, but must be "at least six times" as far in space as the moon.

About the physical nature of the new star, and how it was created, Tycho wisely professed ignorance. Contemporary astronomy calls "new stars" novas, and explains their sudden increase in brightness as an explosive process. There had doubtless been other novas between 125 B.C. and A.D. 1572; but man's new consciousness of the sky, and the new attitude to precise observation, gave the star of 1572 a special significance: the explosion which caused its sudden flaring up shattered the stable, walled-in universe of the ancients.

3. SORCERER'S ISLAND

King Frederick II of Denmark, whose life had been saved by Tycho's foster-father, the late vice-admiral, was a patron of philosophy and the arts. When Tycho was still a student of twenty-four, the king's attention had been called to the brilliant young man, and he had promised him, as a sinecure, the prebend from the first canonry to become vacant. In 1575, when his reputation was already established, Tycho, who loved traveling and did it, like everything else, in the grand style, made a tour of Europe, visiting friends, mostly astronomers, in Frankfurt and Basle, Augsburg, Wittenberg, and Venice, among them the *Landgraf* Wilhelm IV in Cassel. The *Landgraf* was more than an aristocratic dilettante; he had built himself an observatory on a tower in Cassel and was so devoted to astronomy that, when told that his house was on fire while he was observing the new star, he calmly finished his observation before giving his attention to the flames.

He and Tycho got on so well that, after the visit, the *Landgraf* urged King Frederick to provide Tycho with the means for building his own observatory. When Tycho returned to Denmark, Frederick offered him various castles to choose from; but Tycho declined, because he had set his heart on taking up residence in Basle, the charming and civilized old town which had captured the love of Erasmus, Paracelsus, and other illustrious humanists. Now Frederick became really eager to preserve Tycho for Denmark, and in February 1576, sent a messenger—a youth of noble birth, with orders to travel day and night—bearing a royal order for Tycho to come and see the king at once. Tycho obeyed, and the king made him an offer that sounded like a fairy tale: an island in the Sund between Copenhagen and Elsinore Castle, three miles in length, extending over two thousand acres of flat tableland rising on sheer white cliffs out of the sea. Here Tycho should build his house and observatory at Denmark's expense, and in addition receive an annual grant, plus various sinecures, which would make his income one of the highest in Denmark. After a further week's hesitation, Tycho graciously accepted the island of Hveen, and the fortune that went with it.

Accordingly, a royal instrument, signed on May 23, 1576, decreed that:

"We, Frederick the Second, &c., make known to all men, that we of our special favor and grace have conferred and granted in fee, and now by this our open letter confer and grant in fee, to our beloved Tyge Brahe, Otto's son, of Knudstrup, our man and servant, our land of Hveen, with all our and the crown's tenants and servants who thereon live, with all rent and duty which comes from that, and is given to us and to the crown, to have, use, and hold, quit and free, without any rent, all the days of his life, and as long as he lives and likes to continue and follow his *studia mathematices.* . . ."[5]

Thus came into existence the fabulous Uraniburg on the island of Hveen, where Tycho lived for twenty years and taught the world the methods of exact observation.

Tycho's new domain, which he called "the island of Venus, vulgarly named Hveen," had an old tradition of its own. It was often referred to as the "Scarlet Island" —for reasons which a sixteenth-century English traveler explains in his account.

"The Danes think this Island of Wheen to be of such importance, as they have an idle fable, that a King of England should offer for the possession of it, as much scarlet cloth as would cover the same, with a Rose-noble at the corner of each cloth."[6]

It also had some thirteenth-century ruins, to which Danish folklore attached a Niebelung saga all its own. Its inhabitants, distributed over some forty farms grouped around a small village, became the subjects of Tycho, who lorded over them like an Oriental despot.

Tycho's observatory, the Uraniburg, built by a German architect under Tycho's supervision, was a symbol of his character, in which meticulous precision combined with fantastic extravagance. It was a fortress-like monster (see Plate III) which is said to have been "epoch-making in the history of Scandinavian architecture," but on the surviving woodcuts looks rather like a cross between the Palazzo Vecchio and the Kremlin, its Renaissance façade surmounted by an onion-shaped dome, flanked by cylindrical towers, each with a removable top housing Tycho's instruments, and surrounded by galleries with clocks, sundials, globes, and allegorical figures. In the basement were Tycho's private printing press, fed by his own paper mill, his alchemist's furnace and private prison for unruly tenants. He also had his own pharmacy, his game preserves and artificial fish ponds; the only thing he was missing was his tame elk. It had been dispatched to him from his estate but never reached the island. While spending a night in transit at Landskroner Castle, the elk wandered up the stairs to an empty apartment, where it drank so much strong beer that on its way downstairs it stumbled, broke its leg, and died.

In the library stood his largest celestial globe, five feet in diameter, made of brass, on which, in the course of twenty-five years, the fixed stars were engraved one by one, after their correct positions had been newly determined by Tycho and his assistants in the process of remapping the sky; it had cost five thousand dalers, the equivalent of eighty years of Kepler's salary. In the southwest study, the brass arc of Tycho's largest quadrant—fourteen feet in diameter—was fastened to the wall; the space inside the arc was filled with a mural depicting Tycho himself surrounded by his instruments. Later on, Tycho added to the Uraniburg a second observatory, the "Starburg," which was built entirely underground, to protect the instruments from vibration and wind, only the dome-shaped roofs rising above ground level; so that "even from the bowels of the earth he could show the way to the stars and the glory of God."[7] Both buildings were full of gadgets and automata, including statues turning on hidden mechanisms, and a communication system that enabled him to ring a bell in the room of any of his assistants—which made his guests believe that he was convoking them by magic. The guests came in an unceasing procession—savants, courtiers, princes and royalty, including King James VI of Scotland.

Life at Uraniburg was not exactly what one would expect to be the routine of a scholar's family, but rather that of a Renaissance court. There was a steady succession of banquets for distinguished visitors, presided over by the indefatigable, hard-drinking, gargantuan host, holding forth on the variations in the eccentricity of Mars, rubbing ointment on his silver nose, and throwing casual tidbits to his fool, Jepp, who sat at the master's feet under the table, chattering incessantly amidst the general noise. This Jepp was a dwarf, reputed to have second sight, of which he seemed to give spectacular proof on several occasions.

Tycho is really a refreshing exception to the somber, tortured, neurotic geniuses of science. He was, it is true, not a creative genius, only a giant of methodical observation. Still, he displayed all the vanity of genius in his interminable poetic outpourings. His poetry is even more dreadful than Copernicus', and more abundant in quantity—Tycho was never in want of a publisher, since he had his own paper mill and printing press. Even so, his verses and epigrams overflowed onto the murals and ornaments of Uraniburg and Stjoerneburg, which abounded in mottoes, inscriptions, and allegorical figures. The most impressive of these, adorning the wall of his chief study, represented the eight greatest astronomers in history, from Timocharis to Tycho himself, followed by "Tychonides," a yet unborn descendant—with a caption expressing the hope that he would be worthy of his great ancestor.

4. EXILE

Tycho stuck it out on his Scarlet Island for twenty years; then, at fifty-one, he took up his wanderings again. But by that time the bulk of his life's work was done.

In looking back at it, he divided his observations into "childish and doubtful ones" (during his student days at Leipzig), into "juvenile and habitually mediocre ones" (up to his arrival at Hveen), and into "virile, precise, and absolutely certain ones" (made at the Uraniburg).[8] The Tychonic revolution in astronomical method consists in the previously unequaled precision and continuity of his observations. The second point is perhaps even more important than the first: one could almost say that Tycho's work compares with that of earlier astronomers as does a cinematographic record with a collection of still photographs.

In addition to his remarkable survey of the solar system, his remapping of the firmament comprised a thousand fixed stars (of which the positions of 777 were determined accurately, and the remaining 223 places were hastily thrown in just before he left Uraniburg, to make up a round thousand). His proof that the nova of '72 was a true star and that the comet of '77 moved in an orbit far outside the moon's disposed of the already shaken belief in the immutability of the skies and the solidity of the celestial spheres. Last, his system of the world, which he offered as an alternative to the Copernican, though without much scientific value, played, as we shall see, an historically important part.[8a]

The reasons which made Tycho abandon his island realm were of a rather sordid character. Tyge, the Scandinavian squire, was as high-handed in his dealings with men as he was humble toward scientific fact; as arrogant toward his like as he was delicate and tender in handling his instruments. He treated his tenants appallingly, extracting from them labor and goods to which he was not entitled, and imprisoning them when they demurred. He was rude to all who evoked his displeasure, including the young king, Christian IV. The good King Ferdinand had died in 1588 (of too much drink, as Vedel dutifully pointed out in his funeral oration), and his successor, though well disposed to Tycho, on whose sorcerer's island he had spent a delightful day as a boy, was unwilling to close his eyes to Tycho's scandalous rule of Hveen. By this time Tycho's arrogance seemed to be verging on mania of grandeur. He left several letters of the young king unanswered, flouted the decisions of the provincial courts, and even of the high court of justice, by holding a tenant and all his family in chains. As a result, the great man who had been Denmark's glory became a personage thoroughly disliked throughout the country. No direct steps were taken against him, but his fantastic sinecures were reduced to more reasonable proportions, and this gave Tyge, who was becoming increasingly bored and restless on his Scarlet Island, the needed pretext to resume his wanderings again.

He had been preparing his emigration for several years, and when he left Hveen around Easter 1597, he did it in his customary grandiose manner, traveling with a suite of twenty—family, assistants, servants, and the dwarf Jepp—his baggage comprising the printing press, library, furniture, and all the instruments (except the four largest, which followed later). Ever since, as a student, he had ordered his first quadrant at Augsburg, he had been careful to have all his instruments made in such a way that they could be dismantled and transported. "An astronomer," he declared, "must be cosmopolitan, because ignorant statesmen cannot be expected to value their services."[9]

The first station of the Tychonic caravan was Copenhagen, the next Rostock, from where, having left Danish territory, Tycho wrote a rather impertinent letter to King Christian, complaining about the treatment he had received from his ungrateful country, and declaring his intention "to look for help and assistance from other princes and potentates," yet graciously expressing his willingness to return "if it could be done on fair conditions and without injury to myself." Christian wrote back a remarkable letter which soberly refuted Tycho's complaints point by point, and made it clear that the condition of his return to Denmark was "to be respected by you in a different manner if you are to find in us a gracious lord and King."[10]

For once Tyge had found his match. There were only

two men in his life who got the better of him, King Christian of Denmark, and Johannes Kepler from Weilder-Stadt.

His bridges burned, Tycho and his private circus continued their wanderings for another two years—to Wandsbeck Castle near Hamburg, to Dresden, to Wittenberg. Last, in June 1599, they arrived in—or rather made their entry into—Prague, residence of the Emperor Rudolph II, to whom, by the grace of God, Tycho de Brahe had been appointed imperial mathematicus. He was again to have a castle of his choice, and a salary of three thousand florins a year (Kepler in Gratz had two hundred), in addition to some "uncertain income which might amount to some thousands."[11]

Had Tycho remained in Denmark, it is highly unlikely that Kepler could have afforded the expense of visiting him during the short remaining span of Tycho's life. The circumstances which made them both exiles, and guided them toward their meeting, can be attributed to coincidence or providence, according to taste, unless one assumes the existence of some hidden law of gravity in history. After all, gravity in the physical sense is also merely a word for an unknown force acting at a distance.

5. PRELUDE TO THE MEETING

Before they met in the flesh at Benatek Castle, near Prague, Kepler and Tycho had been corresponding for two years.

The relationship had started on the wrong foot, owing to an innocent blunder which young Kepler committed. The episode involved Tycho's lifelong bitter enemy, Ursus the Bear, and makes the fathers of astronomy appear like actors in an *opera buffa*.

Reymers Baer,* who came from Ditmar, had started

---

* German for bear, hence his latinized name, Ursus.

as a swineherd, and ended up as imperial mathematicus
—at which post Tycho was to succeed him, and Kepler
was to succeed Tycho. To achieve, in the sixteenth century, such a career, certainly required considerable gifts
—which, in Ursus, were combined with a dogged and
ferocious character, always ready to crush his victims'
bones in a bearlike hug. In his youth he had published
a Latin grammar and a book on land surveying, then entered the service of a Danish nobleman called Erik
Lange. In 1584, Lange visited Tycho at Uraniburg, and
took Ursus with him. It must have been a rather hectic
encounter, as will presently be seen.

Four years after that visit, Ursus published his *Fundaments of Astronomy*,[12] in which he explained his system of the universe. It was, except for some details, the
same system Tycho had worked out in secret, but had
not published yet, since he wanted more data to elaborate it. In both systems the earth was reinstated as the
center of the world, but the five planets were now cir-
cling round the sun and, with the sun, round the earth.[13]
This was obviously a revival of the intermediary system
between those of Herakleides and Aristarchus of Samos.

Tycho's system was, therefore, by no means very original; but it had the advantage of a compromise between
the Copernican universe and the traditional one. It automatically recommended itself to all those who were reluctant to antagonize academic science, and yet desirous
to "save the phenomena," and was to play an important
part in the Galileo controversy. Actually, the Tychonic
system was "discovered" quite independently by yet a
third scholar, Helisaeus Roeslin, as it so often happens
with inventions that "lie in the air." But Tycho, who was
as proud of his system as Kepler of his five perfect solids,
was convinced that Ursus had stolen it, by snooping
through his manuscripts during that visit in 1584. He
collected evidence to prove that Ursus had been prying
among his papers; that he had taken the precaution of

letting his pupil Andreas share a room with Ursus; that
while Ursus was asleep, the faithful pupil "had taken a
handful of papers out of one of his breeches pockets,
but was afraid to search the other pocket for fear of
waking him"; and that Ursus, on discovering what had
happened, "behaved like a maniac," whereupon all papers which did not concern Tycho were restored to him.

According to Ursus, on the other hand, Tycho had
been haughty and arrogant to him; had tried to shut
him up by remarking that "all these German fellows
are half cracked"; and had been so suspicious about his
observations, "which he was able to take through his
nose, without needing other sights," that he got somebody to search his, Ursus', papers the night before his
departure.

The long and short of it is that the Bear had probably
been snooping among Tycho's observations, but there
is no proof that he had stolen Tycho's "system," nor
that there was any need for him to steal it.

It was into this hornets' nest that young Kepler blundered when he had just hit upon the idea of the *Mysterium* and felt the urgent need to share his joy with the
whole world of learning. Ursus was then the imperial
mathematicus at Prague; so Kepler dashed off a fan letter to him, starting in typical fan-mail style. "There exist
curious men who, unknown, write letters to strangers
in distant lands"; and, continuing with Keplerian effu-
sion, that he was familiar "with the bright glory of thy
fame which makes thee rank first among the *mathematici* of our time like the sun among the minor stars."[14]

This was written in November 1595. The Bear never
answered the unknown young enthusiast's letter; but two
years later, when Kepler was already well known, Ursus
printed the letter, without asking for Kepler's permission, in a book[15] in which he claimed the priority of
the "Tychonic" system, and abused Tycho in most fero-

cious language. The book bore the motto "I will meet them [meaning Tycho and Company] as a bear bereaved of her whelps—Hosea 13:8." Thus Tycho, of course, got the impression that Kepler was siding with the Bear —which was precisely what the Bear had intended. The situation was all the more embarrassing for poor Kepler as he had in the meantime also written a fan letter to Tycho, in which he called him "the prince of mathematicians not only of our time but of all times."[16] Moreover, unaware of the Homeric battle between the two, he had asked Ursus, of all people, to forward a copy of the *Mysterium* to Tycho!

Tycho reacted with unusual diplomacy and restraint. He acknowledged Kepler's letter and book with great courtesy, praised him for the ingenuity of the *Mysterium* while expressing certain reserves, and expressed the hope that Kepler would now make an effort to apply his theory of the five solids to Tycho's own system of the universe. (Kepler wrote on the margin, "Everybody loves himself, but one can see his high opinion of my method.")[17] Only in a postscript did Tycho complain about Kepler's praise of Ursus. A little later Tycho wrote another letter to Maestlin,[18] in which he criticized Kepler's book much more severely, and repeated his complaint. The intention behind this was obvious: Tycho had immediately realized young Kepler's exceptional gifts, wanted to win him over to his side, and hoped that Maestlin would exert his authority with his former pupil in this sense. Maestlin duly transmitted Tycho's complaint to Kepler, who realized only now into what a frightful tangle he had got himself—and, of all people, with Tycho, who was his only hope. So he sat down and penned a long and agonized epistle to Tycho in true Keplerian style, bubbling with sincerity, cheating a little about the facts, pathetic and brilliant and slightly embarrassing, all at the same time.

"How come? Why does he [Ursus] set such value
on my flatteries? . . . If he were a man he would
despise them, if he were wise he would not display
them on the market place. The nonentity which I
then was searched for a famous man who would
praise my new discovery. I begged him for a gift,
and behold, it was he who extorted a gift from the
beggar. . . . My spirit was soaring and melting
away with joy over the discovery I had just made. If,
in the selfish desire to flatter him, I blurted out
words which exceeded my opinion of him, this is to
be explained by the impulsiveness of youth."[19]

And so on. But there is one staggering admission in
the letter: when Kepler read Ursus' *Fundaments of Astronomy* he had believed that the trigonometrical rules
in it were Ursus' original discoveries and did not realize
that most of them could be found in Euclid![20] One
feels the ring of truth in this admission of young Kepler's abysmal ignorance of mathematics at a time when,
guided by intuition alone, he had mapped out the course
of his later achievements in the *Mysterium*.

Tycho replied briefly, and with a gracious condescen-
sion which must have been rather galling to Kepler, that
he had not required such an elaborate apology. Thus the
incident was patched up, though it kept rankling in
Tycho, who, later on, when Kepler became his assistant, would force him to write a pamphlet *In Defense of
Tycho against Ursus*—a chore which Kepler detested.

But for the time being, Tycho was willing to forget the
unfortunate episode, and anxious to get Kepler as his
collaborator. He found it difficult to get the new observatory at Benatek Castle going, and his former assistants were in no hurry to rejoin the former despot of
Hveen. So he wrote to Kepler in December 1599:

"You have no doubt already been told that I have
been most graciously called here by his Imperial

Majesty and that I have been received in the most friendly and benevolent manner. I wish that you would come here, not forced by the adversity of fate, but rather on your own will and desire for common study. But whatever your reason, you will find in me your friend who will not deny you his advice and help in adversity, and will be ready with his help. But if you come soon we shall perhaps find ways and means so that you and your family shall be better looked after in future. *Vale*.

Given at Benatek, or the Venice of Bohemia, on December 9, 1599, by your very sympathetic Tycho Brahe's own hand."[21]

But by the time this letter arrived in Gratz, Kepler was already on his way to Tycho.

he could squeeze half of it out of the Exchequer; when Kepler succeeded him, he would get only a dribble of what was due to him.

By the time Kepler arrived at Benatek, Tycho had already quarreled with the Director of the Crown Estates, who held the purse strings, complained to the Emperor, threatened to leave Bohemia and to tell the world the reasons. Also, several of Tycho's assistants, who had promised to join him at the new Uraniburg, had failed to turn up; and the largest instruments were still delayed on the long trek from Hveen. Toward the end of the year the plague had broken out, obliging Tycho to sit it out with Rudolph at the imperial residence in Girsitz, and supply him with a secret elixir against epidemics. To add to Tycho's worries, Ursus, who had disappeared from Prague on Tycho's arrival, now returned again, trying to create trouble; and Tycho's second daughter, Elisabeth, was having an illicit love affair with one of his assistants, the *Junker* Tengnagel. Young Kepler, in the provincial backwoods of Gratz, had dreamed of Benatek as a serene temple of Urania; he arrived at a madhouse. The castle was teeming with workmen, surveyors, visitors, and the formidable de Brahe clan, including the sinister dwarf Jepp, who huddled under the table during the interminable, tumultuous meals, and found an easy butt for his sarcasms in that timid scarecrow of a provincial mathematicus.

Kepler had arrived in Prague in the middle of January. He had at once written to Benatek, and a few days later received an answer from Tycho, regretting that he could not welcome Kepler in person because of a forthcoming opposition of Mars and Jupiter, to be followed by a lunar eclipse; and inviting him to Benatek "not so much as a guest, than as a very welcome friend and colleague in the contemplation of the skies." The bearers of the letter were Tycho's eldest son and the *Junker* Tengnagel, both of whom were jealous of Kepler from the start, and remained hostile to the end. It was in their company that Kepler completed the last lap of his journey to Tycho— but only after a further delay of nine days. Tengnagel and Tycho Junior were probably having a good time in Prague, and were in no hurry to get back.

At last, then, on February 4, 1600, Tycho de Brahe and Johannes Keplerus, co-founders of a new universe, met face to face, silver nose to scabby cheek. Tycho was fifty-three, Kepler twenty-nine. Tycho was an aristocrat, Kepler a plebeian; Tycho a Croesus, Kepler a church mouse; Tycho a Great Dane, Kepler a mangy mongrel. They were opposites in every respect but one: the irritable, choleric disposition which they shared. The result was constant friction, flaring into heated quarrels, followed by halfhearted reconciliations.

But all this was on the surface. In appearance, it was a meeting of two crafty scholars, each determined to make use of the other for his own purposes. But under the surface, they both knew, with the certainty* of sleepwalkers, that they were born to complete each other; that it was the gravity of fate which had drawn them together. Their relationship was to alternate all the time between these two levels: qua sleepwalkers, they strolled arm in arm through uncharted spaces; in their waking contacts they brought out the worst in the other's character, as if by mutual induction.

Kepler's arrival led to a reorganization of work at Benatek. Previously, Tycho's younger son, Joergen, had been in charge of the laboratory; the senior assistant, Longomontanus, was assigned the study of the orbit of

---

* In the book from which *The Watershed* is taken, Koestler examines the process by which the great discoveries in physical science were made. It is his argument that many of the discoverers failed to recognize the facts even while making correct use of them, much as a sleepwalker might do in his sleep. Hence the title, *The Sleepwalkers*. His thesis is debatable. (Board of Editors)

Mars; and Tycho had intended to put Kepler in charge of the next planet to be taken up for systematic observation. But his eagerness, and the fact that Longomontanus got stuck with Mars, led to a redistribution of the planetary realm among the Tychonites: Kepler was given Mars, the notoriously most difficult planet, while Longomontanus was switched to the moon. This decision proved of momentous importance. Kepler, proud to be entrusted with Mars, boasted that he would solve the problem of its orbit in eight days, and even made a bet with this deadline. The eight days grew into nearly eight years; but out of the struggle of these years with the recalcitrant planet emerged Kepler's *New Astronomy or Physics of the Skies.*

He knew, of course, nothing of what lay ahead of him. He had come to Tycho to wrench from him the exact figures of the eccentricities and mean distances, in order to improve his model of the universe built around the five solids and the musical harmonies. But though he never discarded his *idée fixe*, it was now relegated into the background. The new problems that arose out of Tycho's data "took such a hold of me that I nearly went out of my mind."[2] Himself no more than an amateur observer with the coarsest of instruments, an armchair astronomer with the intuition of genius but still lacking in intellectual discipline, he was overwhelmed by the wealth and precision of Tycho's observations, and only now began to realize what astronomy really meant. The hard facts embodied in Tycho's data, the scrupulousness of Tycho's method, acted like a grindstone on Kepler's fantasy-prone intellect. But although Tycho did the grinding, and the process seemed to be more painful for Kepler than for him, in the end it was the grindstone which was worn down, while the blade emerged sharp and shining from it.

Soon after his arrival in Benatek, Kepler wrote:

Tycho and Kepler

"Tycho possesses the best observations, and thus, so to speak, the material for the building of the new edifice; he also has collaborators and everything he could wish for. He only lacks the architect who would put all this to use according to his own design. For although he has a happy disposition and real architectural skill, he is nevertheless obstructed in his progress by the multitude of the phenomena and by the fact that the truth is deeply hidden in them. Now old age is creeping upon him, enfeebling his spirit and his forces."[3]

There could be no doubt regarding the identity of the architect in Kepler's mind. Nor was it difficult for Tycho to guess Kepler's true opinion of him. He had amassed a treasure of data as nobody before him; but he was old, and lacking the boldness of imagination to build, out of this wealth of raw material, the new model of the universe. Its laws were there, in his columns of figures; but "too deeply hidden" in them for him to decipher. He must also have felt that only Kepler was capable of succeeding in this task—and that nothing could prevent him from succeeding; that it would be this grotesque upstart, and not Tycho himself, nor the hoped-for Tychonides of the Uraniburg mural, who would reap the fruit of his lifelong labors. Half resigned to, half appalled by, his own fate, he wanted at least to make it as difficult for Kepler as possible. He had always been most reluctant to disclose his treasured observations; if Kepler had thought he could simply grab them, he was woefully mistaken—as the indignant complaints in his letters show.

"Tycho gave me no opportunity to share in his experiences. He would only, in the course of a meal, and in between conversing about other matters, mention, as if in passing, today the figure for the apogee of one planet, tomorrow the nodes of another."[4]

One might add: as if he were handing bones to Jepp under the table. Nor would he allow Kepler to copy out his figures. In exasperation, Kepler even asked Tycho's Italian rival, Magini, to offer his own data in exchange for some of Tycho's. Only gradually, step by step, did Tycho yield; and when he put Kepler in charge of Mars, he was forced to disgorge his Mars data.

Kepler had spent barely a month at Benatek when Tycho, in a letter, first hinted at difficulties that had arisen between them; another month later, on April 5, the tension blew up in an explosion which might have shattered the future of cosmology.

The immediate cause of the row was a document which Kepler had drafted, and in which the conditions of his future collaboration with Tycho were laid down in unpleasant detail. If he and his family were to live permanently at Benatek, Tycho must provide them with a detached apartment, because the noise and disorder of the household were having a terrible effect on Kepler's gall, and provoked him to violent outbursts of temper. Next, Tycho must obtain a salary for Kepler from the Emperor, and in the meantime pay him fifty florins a quarter. He must also provide the Keplers with specified quantities of firewood, meat, fish, beer, bread, and wine. As for their collaboration, Tycho must leave Kepler his freedom to choose the time and subject of his work, and only ask him to undertake such researches that were directly connected with it; and since Kepler was "not in need of a spur but rather of a brake to prevent the threat of galloping consumption due to overwork,"[5] he must be allowed to rest in the daytime if he had worked deep into the night. And so on, for several pages.

This document was not meant for Tycho's perusal. Kepler handed it to a guest, a certain Jessenius, professor of medicine in Wittenberg, who was to serve as an intermediary in the negotiations between Tycho and himself. But whether by chance or intrigue, Tycho got hold of the document, which he could hardly regard as flattering to himself. Nevertheless, he took it with that good-humored magnanimity which lived side by side in the Danish *grand seigneur* with jealousy and bullying. He remained a benevolent despot so long as nobody challenged his rule; and, socially, Kepler was so much his inferior that his carping and bickering demands did not affect Tycho as a challenge. One of the reasons for Kepler's bitterness was, incidentally, that he had been assigned an inferior position at the dinner table.

But above all, Tycho needed Kepler, who alone could put his lifework into proper shape. Hence he sat down to negotiate with Kepler in the presence of Jessenius, patiently rubbing ointment on his nose, a paragon of paternal moderation. This attitude grated even more on Kepler's inferiority complex, and he attacked Tycho, in the latter's words, "with the vehemence of a mad dog, to which creature he, Kepler himself, so much likes to compare himself in irritability."[6]

Immediately after the stormy session Tycho, who always had an eye on posterity, wrote down the minutes of it, and requested Jessenius to endorse them. However, when his temper had cooled down, he entreated Kepler to stay on, at least for another few days, until an answer arrived from the Emperor, whom Tycho had approached concerning Kepler's employment. But Kepler refused to listen, and on the next day departed in the company of Jessenius to Prague, where he took quarters with Baron Hoffman. Just before his departure, Kepler had another choleric outburst; at the moment of farewell, he was overcome with remorse and apologized; while Brahe whispered into Jessenius' ear that he should try to bring the *enfant terrible* back to reason. But as soon as they arrived in Prague, Kepler wrote another abusive letter to Tycho.

He must have been in a dreadful state of hysteria. He was suffering from one of his recurrent obscure fevers; his family was in faraway Gratz; the persecution of the Protestants in Styria and the debacle at Benatek had made a shambles of his future; and the data on Mars remained inaccessibly in Tycho's hands. Within a week, the pendulum swung to the other extreme: Kepler wrote a letter of apology to Tycho which sounds like the ravings of a masochist against his own guilty ego.

"The criminal hand which, the other day, was quicker than the wind in inflicting injury, hardly knows how to set about it to make amends. What shall I mention first? My lack of self-control, which I can only remember with the greatest pain, or your benefactions, noblest Tycho, which can neither be enumerated nor valued according to merit? For two months you have most generously provided for my needs . . . you have extended to me every friendliness, you have allowed me to share in your most cherished possession. . . . Taken all in all, neither to your children, nor to your wife, nor to yourself did you devote yourself more than to me. . . . Therefore I think with the deepest dismay that God and the Holy Ghost delivered me to such an extent to my impetuous attacks and to my sick mind that instead of displaying moderation, I indulged during three weeks with closed eyes in sullen stubbornness against you and your family; that instead of thanking you, I displayed blind rage; that instead of showing you respect, I displayed the greatest insolence against your person, which by noble descent, prominent learning, and great fame deserves all respect; that instead of sending you a friendly greeting, I let myself be carried away by suspicion and insinuation when I was itching with bitterness. . . I never considered how cruelly I must have hurt you by this despicable behavior. . . . I come to you as a postulant to ask, in the name of divine pity, for your forgiveness of my terrible offenses. What I have said or written against your person, your fame, your honor, and your scientific rank. . . . I retract in all parts, and declare it voluntarily and freely as invalid, false, and unsound. . . . I also promise sincerely that henceforth at whatever place I shall be I shall not only refrain from such foolish acts, words, deeds, and writings, but I shall also never and in no way unjustly and deliberately offend you. . . . But since the ways of men are slippery, I ask you that whenever you notice in me any tendency toward such unwise manner of behavior, to remind me of myself; you will find me willing. I also promise . . . to oblige you by all kinds of services and . . . thus to prove by my acts that my attitude toward your person is different, and always was different, from what one may conclude from the reckless condition of my heart and body during these last three weeks. I pray that God may help me to fulfill this promise."[7]

I have quoted this letter at some length, because it reveals the tragic core of Kepler's personality. These turns of phrase do not seem to come from a scholar of repute, but from a tortured adolescent, begging to be forgiven by a father whom he hates and loves. Tycho had replaced Maestlin. At the base of his iridescent, complex character, Kepler always remained a waif and stray.

But Tycho was no less dependent on Kepler than Kepler on Tycho. In their worldly contacts, Tycho was the old man of the tribe, Kepler the nagging, ill-mannered adolescent. But on that other level, the rules were reversed: Kepler was the magician from whom, Tycho hoped, would come the solution of his problems, the answer to his frustrations, the salvation from ultimate defeat; and however foolishly they both behaved, they both knew all this.

Therefore, three weeks after the row, Tycho turned up in Prague and drove Kepler back to Benatek in his coach—one can almost see Tyge's great fat arm in the leg-of-mutton sleeve, crushing in an affectionate embrace Kepler's skimpy bones.

2. THE INHERITOR

Altogether the association between Kepler and Tycho lasted for eighteen months, until Tycho's death. Fortunately for both, and for posterity, they were only part of this time in personal contact, for Kepler twice returned to Gratz and spent a total of eight months there to settle his affairs and get his wife's property out.

He left for Gratz the first time shortly after his reconciliation with Tycho, in June 1600. Though peace had been re-established, nothing definite had been settled regarding their future collaboration,[7a] and Kepler was in two minds whether he would return to Tycho or not. He still hoped either to save his position and salary in Gratz by being granted a long leave of absence, or to obtain a chair in his native Wuerttemberg—his lifelong ambition. He wrote to Maestlin and Herwart, his adoptive fathers Numbers One and Two, hinting that Number Three was rather a disappointment; but nothing came of it. He sent to the Archduke Ferdinand a treatise on a solar eclipse, also to no avail; but in that treatise he hit on something for which he had not looked: that there was "a force in the earth" which influenced the moon's motion, a force which diminished in proportion to distance. As he had already attributed a physical force to the sun as an explanation of the motions of the planets, the dependence of the moon on a similar force in the earth was the next important step toward the concept of universal gravity.

But such trifles could not deter the archduke from his plan to stamp out heresy in his lands. On July 31 and the following days, all Lutheran citizens of Gratz, a few more than a thousand in number, had to appear, one by one, before an ecclesiastical commission, and either to declare their willingness to return to the Roman faith, or to suffer expulsion. This time no exemption was made, not even for Kepler—though he was let off paying half the exit levy and granted other financial privileges. The day after he appeared before the commission, a rumor was rife in Gratz that he had changed his mind and declared his readiness to become a Catholic. Whether he had really wavered or not is impossible to know; but in any case, he overcame the temptation and accepted exile with all its consequences.

He sent a last SOS to Maestlin.[8] It starts with a dissertation on the eclipse of the sun on July 10, which he had observed through a *camera obscura* of his own construction, erected in the middle of the market place in Gratz—with the twofold result that a thief stole his purse containing thirty florins, while Kepler himself discovered an important new optical law. The letter continues with the threat that Kepler plus his family would travel down the Danube into Maestlin's arms and a professorship (even if only a small one) which Maestlin would no doubt provide; and ends with the request that Maestlin should pray for him. Maestlin answered that he would gladly pray, but could do nothing else for Kepler, "the steadfast and valiant martyr of God";[9] and after that, answered none of Kepler's letters for four years. He probably thought that he had done his share, and that it was now Tycho's turn to look after the infant prodigy.

Tycho himself was delighted with the sad news. He had doubted whether Kepler would return to him, and

welcomed the prospect all the more as his senior assistant, Longomontanus, had in the meantime left. When
Kepler informed him of his impending expulsion, he
wrote back that Kepler should come at once; "do not
hesitate, make haste, and have confidence."[10] He added
that during a recent audience with the Emperor he had
requested that Kepler should be officially attached to
his observatory, and that the Emperor had nodded his
consent. But in a postscript to the long and affectionate
letter, Tycho could not refrain from alluding to a subject
which had been one of the main reasons for Kepler's un-
happiness at Benatek. On his arrival there, Tycho had
imposed on him the irksome chore of writing a pamphlet
refuting the claims of Ursus; and though Ursus had in
the meantime died, Tycho still insisted on persecuting
him beyond the grave. Moreover, Kepler was also to
write a refutation of a pamphlet by John Craig, physician to James of Scotland, in which Craig had dared to
doubt Tycho's theories about comets. It was not a joyous
prospect for Kepler to waste his time on these futile
labors to serve Tycho's vanity; but now he had no other
choice.

In October he arrived back in Prague with his wife—
but without his furniture and chattels, which he had to
leave behind in Linz, as he had no money to pay for the
transport. He was again ill with intermittent fever, and
again thought that he was suffering from consumption.
The imperial nod of consent to his employment was not
followed by concrete action, so Kepler and his wife had
to live entirely on Tycho's bounty. At the Emperor's request, who wanted his mathematicus close at hand,
Tycho had given up the splendors of Benatek and moved
to a house in Prague, where the Keplers, having no
money for rent, were forced to take up quarters. During
the next six months Kepler had little time for astronomy, as he was fully occupied with writing the accursed
polemics against Ursus and Craig, and nursing his real

and imaginary ailments. Frau Barbara, who even in better days had not been a cheerful soul, hated the alien ways and narrow, winding streets of Prague, whose stench was strong enough "to drive back the Turks," as a contemporary English traveler wrote.[11] The Keplers were drinking the bitter cup of refugee existence to the dregs.

In the spring of 1601, Frau Barbara's rich father died back in Styria—he had paid the price of conversion to die in his country. This gave Kepler a welcome pretext to leave his family in Tycho's charge, and to go back to Gratz to save the inheritance. In this he did not succeed; but he stayed in Gratz for another four months, and seems to have had a wonderful time, dining in the houses of the Styrian nobles as a kind of distinguished exile on home leave, climbing mountains to measure the curvature of the earth, and writing infuriating letters to Tycho, whom he reproached for not giving enough money to Frau Barbara. He returned to Prague in August, his mission unaccomplished, but his health fully restored, and in radiant spirits. He now only had to mark time for another two months till the decisive turn in his life.

On October 13, 1601, Tycho was a guest at supper at Baron Rosenberg's table in Prague. Among the other guests was an imperial councilor, so it must have been an illustrious company; but since Tycho had been in the habit of entertaining royalty, and was accustomed to vast amounts of drink, it is difficult to understand why he was unable to cope with the predicament in which he found himself. Kepler has carefully recorded what happened in the Diary of Observations—a kind of logbook where all important events of the Brahe household were entered.

"On October 13, Tycho Brahe, in the company of Master Minkowitz, had dinner at the illustrious

Rosenberg's table, and held back his water beyond the demands of courtesy. When he drank more, he felt the tension in his bladder increase, but he put politeness before his health. When he got home, he was scarcely able to urinate.

At the beginning of his illness, the moon was in opposition to Saturn . . . [follows the horoscope of the day].

After five sleepless nights, he could still only pass his water with the greatest pain, and even so, the passage was impeded. The insomnia continued, with internal fever gradually leading to delirium; and the food he ate, and from which he could not be kept, exacerbating the evil. On October 24, his delirium ceased for several hours; nature conquered, and he expired peacefully among the consolations, prayers, and tears of his people.

So from this date the series of celestial observations was interrupted, and his own observations of thirty-eight years have come to an end.

On his last night in his gentle delirium, he repeated over and again these words, like someone composing a poem:

Let me not seem to have lived in vain.

No doubt he wished that these words should be added to the title page of his works, thus dedicating them to the memory and uses of posterity."[12]

During his last days, whenever the pain subsided, the great Dane had refused to keep to a diet, ordered and ate ravenously whatever dish came to his mind. When delirium set in again, he kept repeating softly that he hoped his life had not been wasted (*ne frusta vixisse videar*). The meaning of these words becomes clear through his last wish addressed to Kepler.[13] It was the same wish that he had expressed in his first letter to him: that Kepler should build the new universe, not on the Copernican, but on Tycho's system. Yet he must have known, as his delirious complaint revealed, that Kepler would do just the opposite, and put the Tychonic legacy to his own use.

Tycho was buried with great pomp in Prague, his coffin carried by twelve imperial gentlemen-at-arms, preceded by his coat of arms, his golden spurs, and favorite horse.

Two days later, on November 6, 1601, the Emperor's privy councilor, Barwitz, called on Kepler at his lodgings, to appoint him, as Tycho's successor, to the post of imperial mathematicus.

chapter six

The Giving of the Laws

1. "ASTRONOMIA NOVA"

Kepler stayed in Prague as imperial mathematicus from 1601 to 1612, to the death of Rudolph II.

It was the most fruitful period of his life, and brought him the unique distinction of founding two new sciences: instrumental optics, which does not concern us, and physical astronomy. His magnum opus, published in 1609, bears the significant title:

A NEW ASTRONOMY Based on Causation
or A PHYSICS OF THE SKY
derived from Investigations of the
MOTIONS OF THE STAR MARS
Founded on Observations of
THE NOBLE TYCHO BRAHE[1]

Kepler worked on it, with interruptions, from his arrival at Benatek in 1600, to 1606. It contains the first two of Kepler's three planetary Laws: (1) that the planets travel round the sun not in circles but in elliptical orbits, one focus of the ellipse being occupied by the sun; (2) that a planet moves in its orbit not at uniform speed but in such a manner that a line drawn from the planet to the sun always sweeps over equal areas in equal times. The Third Law, published later, does not concern us at this point.

The Giving of the Laws

On the surface, Kepler's Laws look as innocent as Einstein's $E = Mc^2$, which does not reveal, either, its atom-exploding potentialities. But the modern vision of the universe was shaped, more than by any other single discovery, by Newton's law of universal gravitation, which in turn was derived from Kepler's three Laws. Although (owing to the peculiarities of our educational system), a person may never have heard of Kepler's Laws, his thinking has been molded by them without his knowledge; they are the invisible foundation of a whole edifice of thought.

Thus the promulgation of Kepler's Laws is a landmark in history. They were the first "natural laws" in the modern sense: precise, verifiable statements about universal relations governing particular phenomena, expressed in mathematical terms. They divorced astronomy from theology, and married astronomy to physics. Last, they put an end to the nightmare that had haunted cosmology for the last two millennia—the obsession with spheres turning on spheres—and substituted a vision of material bodies not unlike the earth, freely floating in space, moved by physical forces acting on them.

The manner in which Kepler arrived at his new cosmology is fascinating; I shall attempt to retrace the zigzag course of his reasoning. Fortunately, he did not cover up his tracks, as Copernicus, Galileo, and Newton did, who confront us with the result of their labors, and keep us guessing how they arrived at it. Kepler was incapable of exposing his ideas methodically, textbook fashion; he had to describe them in the order they came to him, including all the errors, detours, and the traps into which he had fallen. The New Astronomy is written in an unacademic, bubbling baroque style, personal, intimate, and often exasperating. But it is a unique revelation of the ways in which the creative mind works.

"What matters to me," Kepler explained in his

preface, "is not merely to impart to the reader what I have to say, but above all to convey to him the reasons, subterfuges, and lucky hazards which led me to my discoveries. When Christopher Colombus, Magelhaen, and the Portuguese relate how they went astray on their journeys, we not only forgive them, but would regret to miss their narration because without it the whole, grand entertainment would be lost. Hence I shall not be blamed if, prompted by the same affection for the reader, I follow the same method."[1a]

Before embarking on the story, it will be prudent to add my own apology to Kepler's. Prompted by the same "affection for the reader" I have tried to simplify as far as possible a difficult subject: even so, the present chapter must of necessity be slightly more technical than the rest of this book. If some passages tax his patience, even if occasionally he fails to grasp a point or loses the thread, he will, I hope, nevertheless get a general idea of Kepler's odyssey of thought, which opened up the modern universe.

2. OPENING GAMBITS

It will be remembered that at the partitioning of the cosmos which followed young Kepler's arrival at Benatek Castle, he was allotted the study of the motions of Mars which had defeated Tycho's senior assistant, Longomontanus, and Tycho himself.

"I believe it was an act of Divine Providence," he commented later on, "that I arrived just at the time when Longomontanus was occupied with Mars. For Mars alone enables us to penetrate the secrets of astronomy which otherwise would remain forever hidden from us."[2]

The Giving of the Laws

The reason for this key position of Mars is that, among the outer planets, its orbit deviates more than the others' from the circle; it is the most pronouncedly elliptical. It was precisely for that reason that Mars had defied Tycho and his assistant: since they expected the planets to move in circles, it was impossible to reconcile theory with observation.

> "He [Mars] is the mighty victor over human inquisitiveness, who made a mockery of all the stratagems of astronomers, wrecked their tools, defeated their hosts; thus did he keep the secret of his rule safe throughout all past centuries and pursued his course in unrestrained freedom; wherefore that most famous of Latins, the priest of nature Pliny, specially indicted him: MARS IS A STAR WHO DEFIES OBSERVATION."[3]

Thus Kepler, in his dedication of the *New Astronomy* to the Emperor Rudolph II. The dedication is written in the form of an allegory of Kepler's war against Mars, begun under "Tycho's supreme command," patiently pursued, in spite of the warning example of Rheticus who went off his head over Mars, in spite of other dangers and terrible handicaps, such as a lack of supplies owing to Rudolph's failure to pay Kepler's salary—and so on, to the triumphant end, when the imperial mathematicus, riding a chariot, leads the captive enemy to the Emperor's throne.

Thus Mars held the secret of all planetary motion, and young Kepler was assigned the task of solving it. He first attacked the problem on traditional lines; when he failed, he began to throw out ballast and continued doing so until, by and by, he got rid of the whole load of ancient beliefs on the nature of the universe, and replaced it with a new science.

As a preliminary, he made three revolutionary innovations to gain elbow room, as it were, for tackling his

problem. It will be remembered that the center of Copernicus' system was not the sun, but the center of the
earth's orbit; and that already in the *Mysterium Cosmographicum* Kepler had objected to this assumption as
physically absurd. Since the force that moved the planets emanated from the sun, the whole system should be
centered on the body of the sun itself.[4]

But in fact it was not. The sun occupies not the exact
center of the orbit at C; it occupies one of the two foci
of the ellipse at S.

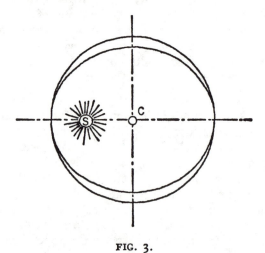

FIG. 3.

Kepler did not know as yet that the orbit was an ellipse; he still regarded it as a circle. But even so, to get
approximately correct results, the center of the circle had
to be placed at C, and not in the sun. Accordingly, the
question arose in his mind: if the force which moves
the planets comes from S, why do they insist on turning
around C? Kepler answered the question by the assumption that each planet was subject to two conflicting influences: *the force of the sun, and a second force located*

in the planet itself. This tug of war caused it now to approach the sun, now to recede from it.

The two forces are, as we know, gravity and inertia. Kepler, as we shall see, never arrived at formulating these concepts. But he prepared the way for Newton by postulating two dynamic forces to explain the eccentricity of the orbits. Before him, the need for a physical explanation was not felt; the phenomenon of eccentricity was merely "saved" by the introduction of an epicycle or eccentric, which made C turn around S. Kepler replaced the fictitious wheels with real forces.

For the same reason, he insisted on treating the sun as the center of his system not only in the physical but in the geometrical sense, by making the distances and positions of the planets relative to the sun (and not relative to the earth or the center C) the basis of his computations. This shift of emphasis, which was more instinctive than logical, became a major factor in his success.

His second innovation is simpler to explain. The orbits of all planets lie very nearly, but not entirely, in the same plane; they form very small angles with each other —rather like adjacent pages of a book which is nearly, but not entirely, closed. The planes of all planets pass, of course, through the sun—a fact which is self-evident to us, but not to pre-Keplerian astronomy. Copernicus, once again misled by his slavish devotion to Ptolemy, had postulated that the plane of the Martian orbit *oscillates in space;* and this oscillation he made to depend on the position of the earth—which, as Kepler remarks, "is no business of Mars." He called this Copernican idea "monstrous" (though it was merely due to Copernicus' complete indifference to physical reality) and set about to prove that the plane in which Mars moves passes through the sun, and does not oscillate, but forms a fixed angle with the plane of the earth's orbit. Here he met, for once, with immediate success. He proved, by several independent methods, all based on the Tychonic observations, that the angle between the planes of Mars and Earth remained always the same, and that it amounted to 1° 50'. He was delighted, and remarked smugly that "the observations took the side of my preconceived ideas, as they often did before."[5]

The third innovation was the most radical. To gain more elbow room, he had to get out of the strait jacket of "uniform motion in perfect circles"—the basic axiom of cosmology from Plato up to Copernicus and Tycho. For the time being, he still let circular motion stand, but he threw out uniform speed. Again he was guided mainly by physical considerations: if the sun ruled the motions, then its force must act more powerfully on the planet when it is close to the source, less powerfully when away from it; hence the planet will move faster or slower, in a manner somehow related to its distance from the sun.

This idea was not only a challenge to antique tradition; it also reversed the original purpose of Copernicus. Copernicus' original motive for embarking on a reform of the Ptolemaic system was his discontent with the fact that, according to Ptolemy, a planet did not move at uniform speed around the center of its orbit, but only around a point at some distance from the center. This point was called the *punctum equans*—the point in space, from which the planet gave the illusion of "equal motion." Copernicus regarded this arrangement as an evasion of the command of uniform motion, abolished Ptolemy's equants, and added, instead, more epicycles to his system. This did not make the planet's *real* motion either circular, or uniform, but each wheel in the imaginary clockwork which was supposed to account for it did turn uniformly—if only in the astronomer's mind.

When Kepler renounced the dogma of uniform motion, he was able to throw out the epicycles which Copernicus had introduced to save it. Instead, he reverted to the equant as an important calculating device.

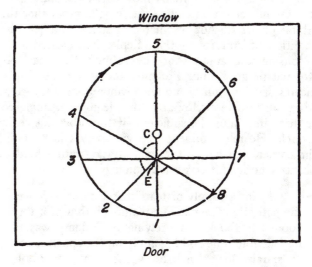

FIG. 4.

Let the circle be the track of a toy train chugging round a room. When near the window, it runs a little faster; near the door, a little slower. Provided that these periodic changes of speed follow some simple, definite rule, then it is possible to find a *punctum equans*, E, from which the train *seems* to move at uniform speed. The closer we are to a moving train, the faster it seems to move; hence the *punctum equans* will be somewhere between the center C of the track and the door, so that the speed surplus of the train when passing the window will be eliminated by distance; its speed deficiency at the door compensated by closeness. The advantage gained by the introduction of this imaginary *punctum equans* is that, *seen from* E, the train seems to move uniformly—that is, it will cover equal angles at equal times—which makes it possible to compute its various positions 1, 2, 3, etc., at any given moment.

By these three preliminary moves: (a) the shifting of the system's center into the sun; (b) the proof that the orbital planes do not "oscillate" in space; and (c) the abolition of uniform motion, Kepler had cleared away a considerable amount of the rubbish that had obstructed progress since Ptolemy, and made the Copernican system so clumsy and unconvincing. In that system Mars ran on five circles; after the clean-up, a single eccentric circle must be sufficient—if the orbit was really a circle. He felt confident that victory was just around the corner, and before the final attack wrote a kind of obituary notice for classical cosmology.

"Oh, for a supply of tears that I may weep over the pathetic diligence of Apianus [author of a very popular textbook] who, relying on Ptolemy, wasted his valuable time and ingenuity on the construction of spirals, loops, helixes, vortices, and a whole labyrinth of convolutions, in order to represent that which exists only in the mind, and which Nature entirely refuses to accept as her likeness. And yet that man has shown us that, with his penetrating intelligence, he would have been capable of mastering Nature."[6]

3. THE FIRST ASSAULT

Kepler's first attack on the problem is described in great detail in the sixteenth chapter of the *New Astronomy*.

The task before him was to define the orbit of Mars by determining the radius of the circle, the direction (relative to the fixed stars) of the axis connecting the two positions where Mars is nearest and farthest from the sun (perihelion and aphelion), and the positions of the sun (S), orbital center (C), and *punctum equans* (E), which all lie on that axis. Ptolemy had assumed that the distance between E and C was the same as between C and S, but Kepler made no such assumption, which complicated his task even more.[7]

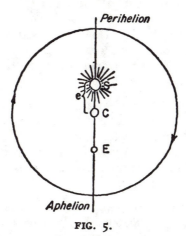

FIG. 5.

He chose out of Tycho's treasure four observed positions of Mars at the convenient dates when the planet was in opposition to the sun.[8] The geometrical problem which he had to solve was, as we saw, to determine, out of these four positions, the radius of the orbit, the direction of the axis, and the position of the three central points on it. It was a problem which could not be solved by rigorous mathematics, only by approximation—that is, by a kind of trial-and-error procedure which has to be continued until all the pieces in the jigsaw puzzle fit together tolerably well. The incredible labor that this involved may be gathered from the fact that Kepler's draft calculations (preserved in manuscript) cover nine hundred folio pages in small handwriting.

At times he was despairing; he felt, like Rheticus, that a demon was knocking his head against the ceiling, with the shout, "These are the motions of Mars!" At other times, he appealed for help to Maestlin (who turned a deaf ear), to the Italian astronomer, Magini (who did the same), and thought of sending an SOS to François Vieta, the father of modern algebra: "Come, O Gallic Apollonius, bring your cylinders and spheres and what other geometer's houseware you have. . . ."[8a] But in the end he had to slog it out alone, and to invent his mathematical tools as he went along.

Halfway through that dramatic sixteenth chapter, he bursts out:

> "If thou [dear reader] art bored with this weari-some method of calculation, take pity on me who had to go through with at least seventy repetitions of it, at a very great loss of time; nor wilst thou be surprised that by now the fifth year is nearly past since I took on Mars. . . ."

Now, at the very beginning of the hair-raising compu-tations in Chapter Sixteen, Kepler absent-mindedly put three erroneous figures for three vital longitudes of Mars, and happily went on from there, never noticing his error. The French historian of astronomy Delambre later re-peated the whole computation, but astonishingly his cor-rect results differ very little from Kepler's faulty ones. The reason is that toward the end of the chapter Kepler committed several mistakes in simple arithmetic—errors in division which would bring bad marks to any school-boy—and these errors happen very nearly to cancel out his earlier mistakes. We shall see, in a moment, that, at the most crucial point of the process of discovering his Second Law, Kepler again committed mathematical sins which mutually canceled out, and "as if by miracle" (in his own words) led to the correct result.

At the end of that breathtaking chapter, Kepler seems to have triumphantly achieved his aim. As a result of his seventy-odd trials, he arrived at values for the radius of the orbit and for the three central points which gave, with a permissible error of less than 2', the correct positions of Mars for all the ten oppositions recorded by Tycho. The unconquerable Mars seemed at last to have been conquered. He proclaimed his victory with the sober words:

"Thou seest now, diligent reader, that the hypothesis based on this method not only satisfies the four positions on which it was based, but also correctly represents, within two minutes, all the other observations. . . ."[9]

There follow three pages of tables to prove the correctness of his claim; and then, without further transition, the next chapter starts with the following words:

"Who would have thought it possible? This hypothesis, which so closely agrees with the observed oppositions, is nevertheless false. . . ."

4. THE EIGHT MINUTES' ARC

In the two following chapters Kepler explains, with great thoroughness and an almost masochistic delight, how he discovered that the hypothesis is false, and why it must be rejected. In order to prove it by a further test, he had selected two specially rare pieces from Tycho's treasury of observations, and lo! they did not fit; and when he tried to adjust his model to them, this made things even worse, for now the observed positions of Mars differed from those which his theory demanded by magnitudes up to eight minutes' arc.

This was a catastrophe. Ptolemy, and even Copernicus, could afford to neglect a difference of eight minutes, because their observations were only accurate within a margin of ten minutes, anyway. "But," the nineteenth chapter concludes, "but for us, who, by divine kindness were given an accurate observer such as Tycho Brahe, for us it is fitting that we should acknowledge this divine gift and put it to use. . . . Henceforth I shall lead the way toward that goal according to my own ideas. For, if I had believed that we could ignore these eight minutes, I would have patched up my hypothesis accordingly. But since it was not permissible to ignore them, those eight minutes point the road to a complete reformation of astronomy: they have become the building material for a large part of this work. . . ."[10]

It was the final capitulation of an adventurous mind before the "irreducible, obstinate facts." Earlier, if a minor detail did not fit into a major hypothesis, it was cheated away or shrugged away. Now this time-hallowed indulgence had ceased to be permissible. A new era had begun in the history of thought: an era of austerity and rigor. As Alfred North Whitehead put it:

"All the world over and at all times there have been practical men, absorbed in 'irreducible and stubborn facts': all the world over and at all times there have been men of philosophic temperament who have been absorbed in the weaving of general principles. It is this union of passionate interest in the detailed facts with equal devotion to abstract generalization which forms the novelty in our present society."[11]

This new departure determined the climate of European thought in the last three centuries; it set modern Europe apart from all other civilizations in the past and present, and enabled it to transform its natural and social environment as completely as if a new species had arisen on this planet.

The turning point is dramatically expressed in Kepler's work. In the *Mysterium Cosmographicum* the facts are coerced to fit the theory. In the *Astronomia Nova*, a theory built on years of labor and torment was instantly thrown away because of a discord of eight miserable minutes' arc. Instead of cursing those eight minutes as a stumbling block, he transformed them into the cornerstone of a new science.

What caused this change of heart in him? It is easy to find some of the general causes which contributed to the emergence of the new attitude: the need of navigators, and engineers, for greater precision in tools and theories; the stimulating effects on science of expanding commerce and industry. But what turned Kepler into the first lawmaker of nature was something different and more specific. It was *his introduction of physical causality into the formal geometry of the skies* which made it impossible for him to ignore the eight minutes' arc. So long as cosmology was guided by purely geometrical rules of the game, regardless of physical causes, discrepancies between theory and fact could be overcome by inserting another wheel into the system. In a universe moved by real, physical forces, this was no longer possible. The revolution that freed thought from the stranglehold of ancient dogma immediately created its own rigorous discipline.

Book Two of the *New Astronomy* closes with the words:

> "And thus the edifice which we erected on the foundation of Tycho's observations, we have now again destroyed. . . . This was our punishment for having followed some plausible, but in reality false, axioms of the great men of the past."

5. THE WRONG LAW

The next act of the drama opens with Book Three. As the curtain rises, we see Kepler preparing himself to throw out more ballast. The axiom of *uniform* motion has already gone overboard; Kepler feels, and hints,[12] that the even more sacred one of *circular* motion must follow. The impossibility of constructing a circular orbit that would satisfy all existing observations suggests to him that the circle must be replaced with some other geometrical curve.

But before he can do that, he must make an immense detour. For if the orbit of Mars is not a circle, its true shape can be discovered only by defining a sufficient number of points on the unknown curve. A circle is defined by three points on its circumference; every other curve needs more. The task before Kepler was to construct Mars' orbit without any preconceived ideas regarding its shape—to start from scratch, as it were.

To do that, it was first of all necessary to re-examine the motion of the earth itself. For, after all, the earth is our observatory; and if there is some misconception regarding its motion, all conclusions about the motions of other bodies will be distorted. Copernicus had assumed that the earth moves at uniform speed—not, as the other planets, only "quasi-uniformly" relative to some equant or epicycle, but *really* so. And since observation contradicted the dogma, the inequality of the earth's motion was explained away by the suggestion that the orbit periodically expanded and contracted, like a kind of pulsating jellyfish.[13] It was typical of the improvisations that astronomers could afford so long as they felt free to manipulate the universe on their drawing boards as they pleased. It was equally typical that Kepler rejected it as "fantastic,"[14] again on the grounds that no physical cause existed for such a pulsation.

Hence his next task was to determine, more precisely than Copernicus had done, the earth's motion around the sun. For that purpose he designed a highly original method of his own. It was relatively simple, but it so happened that nobody had thought of it before. It consisted, essentially, in the trick of transferring the observer's position from Earth to Mars, and to compute the motions of the earth exactly as an astronomer on Mars would do it.[15]

136

The result was just as he had expected: the earth, like the other planets, did not revolve with uniform speed, but faster or slower according to its distance from the sun. Moreover, at the two extreme points of the orbit, the aphelion and perihelion (see Fig. 5) the earth's velocity proved to be, simply and beautifully, inversely proportional to distance.

At this decisive point,[16] Kepler flies off the tangent and becomes airborne, as it were. Up to here he was preparing, with painstaking patience, his second assault on the orbit of Mars. Now he turns to a quite different subject. "Ye physicists, prick your ears," he warns, "for now we are going to invade your territory."[17] The next six chapters are a report on that invasion into celestial physics, which had been out of bounds for astronomy since Plato.

A phrase seems to have been humming in his ear like a tune one cannot get rid of; it crops up in his writings over and again: there is a force in the sun which moves the planet, there is a force in the sun, there is a force in the sun. And since there is a force in the sun, there must exist some beautifully simple relation between the planet's distance from the sun and its speed. A light shines the brighter, the nearer we are to its source, and the same must apply to the force of the sun: the closer the planet to it, the quicker it will move. This is his instinctive conviction, already expressed in the *Mysterium Cosmographicum*; but now, at last, he has succeeded in proving it.

In fact he has not. He has proved the inverse ratio of speed to distance only for the *two extreme points* of the orbit; and his extension of this "Law" to the *entire* orbit was a patently incorrect generalization. Moreover, Kepler knew this, and admitted it at the end of the thirty-second chapter,[18] before he became airborne; but immediately afterward, he conveniently forgot it. This is the first of the critical mistakes which "as if by a miracle" canceled out, and led Kepler to the discovery of his Second Law. It looks as if his conscious, critical faculties were anesthetized by the creative impulse, by his impatience to get to grips with the physical forces in the solar system.

Since he had no notion of the *momentum* that makes the planet persist in its motion, and only a vague intuition of *gravity*, which bends that motion into a closed orbit, he had to find, or invent, a force that, like a broom, sweeps the planet around its path. And since the sun causes all motion, he let the sun handle the broom. This required that the sun rotate round its own axis—a guess which was only confirmed much later; the force which it emitted rotated with it, like the spokes of a wheel, and swept the planets along. But if that were the only force acting on them, the planets would all have the same angular velocity, they would all complete their revolutions in the same period—but they do not. The reason, Kepler thought, was the laziness or "inertia" of the planets, which desire to remain in the same place, and resist the sweeping force. The "spokes" of that force are not rigid; they allow the planet to lag behind; it works rather like a vortex or whirlpool.[19] The power of the whirlpool diminishes with distance, so that the farther away the planet, the less power the sun has to overcome its laziness, and the slower its motion will be.

It still remained to be explained, however, why the planets moved in eccentric orbits instead of always keeping the same distance from the center of the vortex. Kepler first assumed that, apart from being lazy, they performed an epicyclic motion in the opposite direction under their own steam, as it were, apparently out of sheer cussedness. But he was dissatisfied with this, and at a later stage assumed that the planets were "huge round magnets" whose magnetic axis pointed always in the same direction, like the axis of a top; hence the planet will periodically be drawn closer to, and be repelled by the sun, according to which of its magnetic poles faces the sun.

Thus, in Kepler's physics of the universe, the roles played by gravity and inertia are reversed. Moreover, he assumed that the sun's power diminishes in direct ratio to distance. He sensed that there was something wrong here, since he knew that the intensity of light diminishes with the *square* of distance; but he had to stick to it, to satisfy his theorem of the ratio of speed to distance, which was equally false.

6. THE SECOND LAW

Refreshed by this excursion into celestial physics, our hero returned to the more immediate task in hand. Since the earth no longer moved at uniform speed, how could one predict its position at a given time? (The method based on the *punctum equans* had proved, after all, a disappointment.) If its speed depended directly on its distance from the sun, the time it needed to cover a small fraction of the orbit was always proportionate to that distance. Hence he divided the orbit (which, forgetting his previous resolve, he still regarded as a circle) into 360 parts, and computed the distance of each bit of arc from the sun. The sum of all distances between, say 0° and 85°, was a measure of the time the planet needed to get there.

But this procedure was, as he remarked with unusual modesty, "mechanical and tiresome." So he searched for something simpler.

"Since I was aware that there exists an infinite number of points on the orbit and accordingly an infinite number of distances [from the sun] the idea occurred to me that the sum of these distances is contained in the *area* of the orbit. For I remembered that in the same manner Archimedes too divided the area of a circle into an infinite number of triangles."[20]

Accordingly, he concluded, the area swept over by the line connecting planet and sun AS–BS is a measure of the time required by the planet to get from A to B; *hence the line will sweep out equal areas in equal times.* This is Kepler's immortal Second Law (which he discovered before the First)—a law of amazing simplicity at the end of a dreadfully confusing labyrinth.

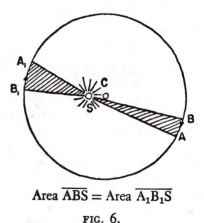

Area $\overline{ABS}$ = Area $\overline{A_1B_1S}$

FIG. 6.

Yet the last step which had got him out of the labyrinth had once again been a faulty step. For it is not permissible to equate an area with the sum of an infinite number of neighboring lines, as Kepler did. Moreover, he knew this well, and explained at length why it was not permissible.[21] He added that he had also committed a second error, by assuming the orbit to be circular. And he concluded: "But these two errors—it is like a miracle—cancel out in the most precise manner, as I shall prove further down."[22]

The correct result is even more miraculous than Kepler realized, for his explanation of the reasons *why* his errors cancel out was once again mistaken, and he got, in fact, so hopelessly confused that the argument is practically impossible to follow—as he himself admitted. And yet, by three incorrect steps and their even more incorrect defense, Kepler stumbled on the correct law.[23] It is perhaps the most amazing sleepwalking performance in the history of science—except for the manner in which he found his First Law, to which we now turn.

7. THE FIRST LAW

The Second Law determined the variations of the planet's speed along its orbit, but it did not determine the shape of the orbit itself.

At the end of the Second Book, Kepler had acknowledged defeat in his attempts to define the Martian orbit —a defeat caused by a discrepancy of eight minutes' arc. He had then embarked on an enormous detour, starting with the revision of the earth's motion, followed by physical speculations, and terminating in the discovery of the Second Law. In the Fourth Book he resumed his investigation of the Martian orbit where he had left off. By this time, four years after his first, frustrated attempts, he had become even more skeptical of orthodox dogma, and had gained an unparalleled skill in geometry by the invention of methods all his own.

The final assault took nearly two years; it occupies Chapters Forty-one to Sixty of the *New Astronomy*. In the first four (forty-one to forty-four), Kepler tried for the last time, with savage thoroughness, to attribute a circular orbit to Mars and failed: this section ends with the words:

"The conclusion is quite simply that the planet's path is not a circle—it curves inward on both sides and outward again at opposite ends. Such a curve is called an oval. The orbit is not a circle, but an oval figure."

But now a dreadful thing happened, and the next six chapters (forty-five to fifty) are a nightmare journey through another labyrinth. This oval orbit is a wild, frightening new departure for him. To be fed up with cycles and epicycles, to mock the slavish imitators of Aristotle is one thing; to assign an entirely new, lopsided, implausible path for the heavenly bodies is quite another.

Why indeed an oval? There is something in the perfect symmetry of spheres and circles which has a deep, reassuring appeal to the unconscious—otherwise it could not have survived two millennia. The oval lacks all such archetypal appeal. It has an arbitrary form. It distorts that eternal dream of the harmony of the spheres, which lay at the origin of the whole quest. Who art thou, Johann Kepler, to destroy divine symmetry? All he has to say in his own defense is that, having cleared the stable of astronomy of cycles and spirals, he left behind him "only a single cartful of dung": his oval.[24]

At this point, the sleepwalker's intuition failed him; he seems to be overcome by dizziness, and clutches at the first prop that he can find. He must find a physical cause, a cosmic *raison d'être* for his oval in the sky—and he falls back on the old quack remedy which he has just abjured, the conjuring up of an epicycle! To be sure, it is an epicycle with a difference: it has a physical cause. We have heard earlier that while the sun's force sweeps the planet around in a circle, a second, antagonistic force, "seated in the planet itself," makes it turn in a small epicycle in the opposite direction. It all seems to him "wonderfully plausible,"[25] for the result of the combined movement is indeed an oval. But a very special oval: it has the shape of an egg, with the pointed end at the perihelion, the broad end at the aphelion.

No philosopher had laid such a monstrous egg before. Or, in Kepler's own words of wistful hindsight:

"What happened to me confirms the old proverb: a bitch in a hurry produces blind pups. . . . But I simply could not think of any other means of imposing an oval path on the planets. When these ideas fell upon me, I had already celebrated my new triumph over Mars without being disturbed by the question . . . whether the figures tally or not. . . . Thus I got myself into a new labyrinth. . . . The reader must show tolerance to my gullibility."[26]

The battle with the egg goes on for six chapters, and took a full year of Kepler's life. It was a difficult year; he had no money, and was down with "a fever from the gall." A threatening new star, the nova of 1604, had appeared in the sky; Frau Barbara was also ill, and gave birth to a son—which provided Kepler with an opportunity for one of his dreadful jokes: "Just when I was busy squaring my oval, an unwelcome guest entered my house through a secret doorway to disturb me."[27]

To find the area of his egg, he again computed a series of 180 Sun-Mars distances and added them together; and this whole operation he repeated no less than forty times. To make the worthless hypothesis work, he temporarily repudiated his own, immortal Second Law—to no avail. Finally, a kind of snow-blindness seemed to descend on him: he held the solution in his hand without seeing it. On July 4, 1603, he wrote to a friend that he was unable to solve the geometrical problems of his egg; but "if only the shape were a perfect ellipse all the answers could be found in Archimedes' and Apollonius' work."[28] A full eighteen months later he again wrote to the same correspondent that the truth must lie somewhere halfway between egg shape and circular shape—"just as if the Martian orbit were a perfect ellipse. But regarding that, I have so far explored nothing."[29] What is even more astonishing, he constantly used ellipses in his calculations—but merely as an *auxiliary* device to determine, by approximation, the area of his egg curve—which by now had become a veritable fixation. Was there some unconscious biological bias behind it? Apart from the association between the squaring of the egg and the birth of his child, there is nothing to substantiate that attractive hypothesis.*

And yet, these years of wandering through the wilderness were not entirely wasted. The otherwise sterile chapters in the *New Astronomy* devoted to the egg hypothesis represent an important further step toward the invention of the infinitesimal calculus. Besides, Kepler's mind had by now become so saturated with the numerical data of the Martian orbit that when the crucial hazard presented itself, it responded at once as a charged cloud reacts to a spark.

This hazard is perhaps the most unlikely incident in this unlikely story. It presented itself in the shape of a number which had stuck in Kepler's brain. The number was 0.00429.

When he at last realized that his egg had "gone up in smoke"[30] and that Mars, whom he had believed a conquered prisoner "securely chained by my equations, immured in my tables," had again broken loose, Kepler decided to start once again from scratch.

He computed very thoroughly a score of Mars-Sun distances at various points of the orbit. They showed again that the orbit was some kind of oval, looking like a circle flattened inward at two opposite sides, so that there were two narrow sickles or lunulae left between the circle and

* Copernicus, too, stumbled on the ellipse and kicked it aside; but Copernicus, who firmly believed in circles, had much less reason to pay attention to it than Kepler, who had progressed to the oval.

the Martian orbit. The width of the sickle, where it is thickest, amounted to 0.00429 of the radius.

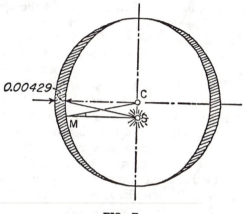

FIG. 7.

At this point Kepler, for no particular reason, became interested in the angle at M—the angle formed between the sun and the center of the orbit, as seen from Mars. This angle was called the "optical equation." It varies, of course, as Mars moves along its orbit; its maximum value is 5° 18'. This is what happened next, in Kepler's own words.[31]

". . . I was wondering why and how a sickle of just this thickness (0.00429) came into being. While this thought was driving me around, while I was considering again and again that . . . my apparent triumph over Mars had been in vain, I stumbled entirely by chance on the secant* of the angle 5° 18', which is the measure of the greatest optical equation. When I realized that this secant equals 1.00429, I felt as if I had been awakened from a sleep. . . ."

* The "secant" of the angle at M is the ratio MC : MS.

It had been a true sleepwalker's performance. At the first moment, the reappearance of the number 0.00429 in this unexpected context must have seemed a miracle to Kepler. But he realized in a flash that the apparent miracle must be due to a fixed relation between the angle at M and the distance to S, a relation which must hold true for any point of the orbit; only the manner in which he had stumbled on that relation was due to chance. "The roads that lead man to knowledge are as wondrous as that knowledge itself."

Now at last, at long last, after six years of incredible labor, he held the secret of the Martian orbit. He was able to express the manner in which the planet's distance from the sun varied with its position, in a simple formula, a mathematical law of nature. *But he still did not realize that this formula specifically defined the orbit as an ellipse.** Nowadays, a student with a little knowledge of analytical geometry would realize this at a glance; but analytical geometry came after Kepler. He had discovered his magic equation empirically, but he could no more identify it as the shorthand sign for an ellipse than the average reader of this book can; it was nearly as meaningless to him. He had reached his goal, but he did not realize that he had reached it.

The result was that he went off on one more last wild-goose chase. He tried to construct the orbit corresponding to his newly discovered equation; but he did not know how, made a mistake in geometry, and arrived at a curve which was too bulgy; the orbit was a *via buccosa*, chubby-faced, as he noted with disgust.

What next? We have reached the climax of the comedy. In his despair, Kepler threw out his formula (which denoted an elliptic orbit) because he wanted to try out an entirely new hypothesis—to wit, an elliptic orbit. It was as if the tourist had told the waiter, after

* In modern denotation, the formula is: $R = 1 + e \cos \beta$ where R is the distance from the sun, β the longitude referred to the center of the orbit, and e the eccentricity.

146

studying the menu: "I don't want *côtelette d'agneau,*
whatever that is; I want a lamb chop."

By now he was convinced that the orbit must be an
ellipse, because countless observed positions of Mars,
which he knew almost by heart, irresistibly pointed to
that curve; but he still did not realize that his equation,
which he had found by chance plus intuition, *was* an
ellipse. So he discarded that equation, and constructed
an ellipse by a different geometrical method. And then,
at long last, he realized that the two methods produced
the same result.

With his usual disarming frankness, he confessed what
had happened.

> "Why should I mince my words? The truth of
> Nature, which I had rejected and chased away, re-
> turned by stealth through the back door, disguising
> itself to be accepted. That is to say, I laid [the orig-
> inal equation] aside, and fell back on ellipses, be-
> lieving that this was a quite different hypothesis,
> whereas the two, as I shall prove in the next chap-
> ter, are one and the same . . . I thought and
> searched, until I went nearly mad, for a reason why
> the planet preferred an elliptical orbit [to mine].
> . . . Ah, what a foolish bird I have been!"[32]

But in the list of contents, in which he gives a brief
outline of the whole work, Kepler sums up the matter in
a single sentence:

> "I show [in this chapter] how I unconsciously re-
> pair my error."

The remainder of the book is in the nature of a
mopping-up operation after the final victory.

8. SOME CONCLUSIONS

It was indeed a tremendous victory. The great Ferris
wheel of human delusion, with its celestial catwalks for
the wandering planets, this phantasmagoria which had

blocked man's approach to nature for two thousand years, was destroyed, "banished to the lumber room." Some of the greatest discoveries consist mainly in the clearing away of psychological road blocks which obstruct the approach to reality; that is why, *post factum*, they appear so obvious. In a letter to Longomontanus[33] Kepler qualified his own achievement as the "cleansing of the Augean stables."

But Kepler not only destroyed the antique edifice; he erected a new one in its place. His Laws are not of the type that appears self-evident, even in retrospect (as, say, the Law of Inertia appears to us); the elliptic orbits and the equations governing planetary velocities strike us as "constructions" rather than "dis-coveries." In fact, they make sense only in the light of Newtonian mechanics. From Kepler's point of view, they did not make much sense; he saw no logical reason why the orbit should be an ellipse instead of an egg. Accordingly, he was more proud of his five perfect solids than of his Laws; and his contemporaries, including Galileo, were equally incapable of recognizing their significance. The Keplerian discoveries were not of the kind that is "in the air" of a period, and usually made by several people independently; they were quite exceptional one-man achievements. That is why the way he arrived at them is particularly interesting.

I have tried to retrace the tortuous progress of his thought. Perhaps the most astonishing thing about it is the mixture of the clean and the unclean in his method. On the one hand, he throws away a cherished theory, the result of years of labor, because of those wretched eight minutes of arc. On the other hand he makes impermissible generalizations, knows that they are impermissible, yet does not care. And he has a philosophical justification for both attitudes. We heard him sermonizing about the duty to stick rigorously to observed fact. But on the other hand he says that Copernicus "sets an example for others by his contempt for the small blemishes in expounding his wonderful discoveries. If this had not been always the usage, then Ptolemy would never have been able to publish his *Almagest*, Copernicus his *Revolutions*, and Reinhold his *Prutenian Tables*. . . . It is not surprising that, when he dissects the universe with a lancet, various matters emerge only in a rough manner."[34]

Both precepts have, of course, their uses. The problem is to know when to follow one, when the other. Copernicus had a one-track mind; he never flew off at a tangent; even his cheatings were heavy-handed. Tycho was a giant as an observer, but nothing else. His leanings toward alchemy and astrology never fused, as in Kepler, with his science. The measure of Kepler's genius is the intensity of his contradictions, and the use he made of them. We saw him plod, with infinite patience, along dreary stretches of trial-and-error procedure, then suddenly become airborne when a lucky guess or hazard presented him with an opportunity. What enabled him to recognize instantly his chance when the number 0.00429 turned up in an unexpected context was the fact that not only his waking mind, but his sleepwalking unconscious self was saturated with every conceivable aspect of his problem, not only with the numerical data and ratios, but also with an intuitive "feel" of the physical forces, and of the *Gestalt* configurations which it involved. A locksmith who opens a complicated lock with a crude piece of bent wire is not guided by logic, but by the unconscious residue of countless past experiences with locks, which lend his touch a wisdom that his reason does not possess. It is perhaps this intermittent flicker of an over-all vision which accounts for the mutually compensatory nature of Kepler's mistakes, as if some balancing reflex or "back-feed" mechanism had been at work in his unconscious mind.

Thus, for instance, he *knew* that his inverse-ratio

"law" (between a planet's speed and solar distance) was incorrect. His thirty-second chapter ends with a short, almost offhand admission of this. But, he argues, the deviation is so small that it can be neglected. Now this is true for Earth, with its small eccentricity, yet not at all true for Mars, with its large eccentricity. Yet even toward the end of the book (in Chapter Sixty), long after he had found the correct law, Kepler speaks of the inverse-ratio postulate as if it were true not only for Earth, but also for Mars. He could not deny, even to himself, that the hypothesis was incorrect; he could only forget it. Which he promptly did. Why? Because, though he knew that the postulate was bad geometry, it made good physics to him, and therefore ought to be true. The problem of the planetary orbits had been hopelessly bogged down in its purely geometrical frame of reference, and when Kepler realized that he could not get it unstuck, he tore it out of that frame and removed it into the field of physics. This operation of removing a problem from its traditional context and placing it into a new one, looking at it through glasses of a different color, as it were, has always seemed to me of the very essence of the creative process.[35] It leads not only to a revaluation of the problem itself, but often to a synthesis of much wider consequences, brought about by a fusion of the two previously unrelated frames of reference. In our case, the orbit of Mars became the unifying link between the two formerly separate realms of physics and cosmology.

It may be objected that Kepler's ideas of physics were so primitive that they ought to be regarded merely as a subjective stimulus to his work (like the five perfect solids), without objective value. In fact, however, his was the first serious attempt at explaining the mechanism of the solar system in terms of physical forces, and once the example was set, physics and cosmology could never again be divorced. And, second, whereas the five solids were indeed merely a psychological spur, his sky physics played, as we have seen, a direct part in the discovery of his Laws.

For, although the functions of gravity and inertia are reversed in the Keplerian cosmos, his intuition that there are *two antagonistic forces* acting on the planets guided him in the right direction. A single force, as previously assumed (that of the Prime Mover or kindred spirits) could never produce oval orbits and periodic changes of speed. These could only be the result of some dynamic tug of war going on in the sky—as indeed they are; though his ideas about the nature of the "sun's force" and the planet's "laziness" or "magnetism" were pre-Newtonian.

9. THE PITFALLS OF GRAVITY

I have tried to show that without his invasion into the territory of physics Kepler could not have succeeded. I must now discuss briefly Kepler's particular brand of physics. It was, as to be expected, physics on the watershed, halfway between Aristotle and Newton. The essential concept of impetus or momentum, which makes a moving body persist in its motion without the help of an external force, is absent from it; the planets must still be dragged through the ether like a Greek oxcart through the mud. In this respect Kepler had not advanced further than Copernicus, and both were unaware of the progress made by the Ockhamists in Paris.

On the other hand, he came very near to discovering universal gravity, and the reasons for his failure to do so are not only of historical but also of topical interest. Over and again he seems to balance on the brink of the idea and yet, as if pulled back by some unconscious resistance, to shrink from the final step. One of the most striking passages is to be found in the introduction to the *Astronomia Nova*. There Kepler starts by demolishing the Aristotelian doctrine that bodies that are by nature "heavy" strive toward the center of the world, whereas those that are "light" strive toward its periphery. His conclusions are as follows.

"It is therefore clear that the traditional doctrine about gravity is erroneous. . . . Gravity is the mutual bodily tendency between cognate [i.e., material] bodies toward unity or contact (of which kind the magnetic force also is), so that the earth draws a stone much more than the stone draws the earth. . . .

Supposing that the earth were in the center of the world, heavy bodies would be attracted to it, not because it is in the center, but because it is a cognate [material] body. It follows that regardless where we place the earth . . . heavy bodies will always seek it. . . .

If two stones were placed anywhere in space near to each other, and outside the reach of force of a third cognate body, then they would come together, after the manner of magnetic bodies, at an intermediate point, each approaching the other in proportion to the other's mass [my italics].

If the earth and the moon were not kept in their respective orbits by a spiritual or some other equivalent force, the earth would ascend toward the moon $\frac{1}{54}$ part of the distance, and the moon would descend the remaining 53 parts of the interval, and thus they would unite. But the calculation presupposes that both bodies are of the same density.

If the earth ceased to attract the waters of the sea, the seas would rise and flow into the moon. . . .

If the attractive force of the moon reaches down to the earth, it follows that the attractive force of the earth, all the more, extends to the moon and even farther. . . .

Nothing made of earthy substance is absolutely light; but matter which is less dense, either by nature or through heat, is relatively lighter. . . .

Out of the definition of lightness follows its motion; for one should not believe that when lifted up, it escapes to the periphery of the world, or that it is not attracted by the earth. It is merely less attracted than heavier matter, and is therefore displaced by heavier matter, so that it comes to rest and is kept in its place by the earth. . . ."[35a]

In the same passage, Kepler gives the first correct explanation of the tides as a motion of the waters "toward the regions where the moon stands in the zenith." In a later work (the *Somnium*) he explained the tides, not by the attraction of the moon alone, but of moon and sun combined; he thus realized that the attraction of the sun reached as far as the earth!

Yet in spite of this, the sun in his cosmology is not an attracting force, but acts like a sweeping broom. In the text of the *New Astronomy* he seems to have forgotten all that he said in the preface about the mutual attraction between two bodies in empty space, and his strikingly correct definition of gravity being proportional to the attracting mass. The definitions of gravity in the preface are indeed so striking that Delambre exclaims,[36] "*Voilà qui était neuf, vraiment beau, et qui n'avait besoin que de quelques developpements et que de quelques explications. Voilà les fondaments de la Physique moderne, céleste et terrestre.*"* But when he tried to work out the mechanics of the solar system, all these beautiful new insights were lost again in confusion. Could some similar paradox be responsible for the crisis

* "Here was something new and truly beautiful, which only needed a little development and explanation. Here were the foundations of modern physics, both of the earth and the skies."

in modern physics—some unconscious blockage which prevents us from seeing the "obvious," and compels us to persist in our own version of wave-mechanical double-think?

At any rate, most twentieth-century physicists will feel a sneaking sympathy for the man who nibbled at the concept of gravity and yet was unable to swallow it. For Newton's concept of a "gravitational force" has always lain as an undigested lump in the stomach of science; and Einstein's surgical operation, though easing the symptoms, has brought no real remedy. The first to sympathize with Kepler would have been Newton himself, who, in a famous letter to Bentley, wrote:

"It is inconceivable, that inanimate brute matter should, without the mediation of something else, which is not material, operate upon, and affect other matter without mutual contact; as it must do, if gravitation, in the sense of Epicurus, be essential and inherent in it. And this is one reason, why I desired you would not ascribe innate gravity to me. That gravity should be innate, inherent, and essential to matter, so that one body may act upon another, at a distance through a vacuum, without the mediation of anything else, by and through which their action and force may be conveyed from one to another, is to me so great an absurdity, that I believe no man who has in philosophical matters a competent faculty of thinking, can ever fall into it."[37]

Newton, in fact, could get over the "absurdity" of his own concept only by invoking either an ubiquitous ether (whose attributes were equally paradoxical) and/or God in person. The whole notion of a "force" that acts instantly at a distance without an intermediary agent, that traverses the vastest distances in zero seconds and pulls at immense stellar objects with ubiquitous ghost fingers

—the whole idea is so mystical and "unscientific" that "modern" minds like Kepler, Galileo, and Descartes, who were fighting to break loose from Aristotelian animism, would instinctively tend to reject it as a relapse into the past.[38] In their eyes, the idea of "universal gravity" would amount to much the same kind of thing as the anima mundi of the ancients. What made Newton's postulate nevertheless a modern law of nature was his mathematical formulation of the mysterious entity to which it referred. And that formulation, Newton deduced from the discoveries of Kepler—who had intuitively glimpsed gravity, and shied away from it. In such crooked ways does the tree of science grow.

10. MATTER AND MIND

In a letter to Herwart, which he wrote when the book was nearing completion,[39] Kepler defined his program.

> "My aim is to show that the heavenly machine is not a kind of divine, live being, but a kind of clockwork (and he who believes that a clock has a soul, attributes the maker's glory to the work), insofar as nearly all the manifold motions are caused by a most simple, magnetic, and material force, just as all motions of the clock are caused by a simple weight. And I also show how these physical causes are to be given numerical and geometrical expression."

He had defined the essence of the scientific revolution. But he himself never completed the transition from a universe animated by purposeful intelligence to one moved by inanimate, "blind" forces. The very concept of a physical "force" devoid of purpose, which we take so much for granted, was only just emerging from the womb of animism, and the word for it—*virtus* or *vis*—betrays its origin. It was (and is) indeed much easier to talk about a "simple, magnetic, material force" than to form a concrete idea of its working. The following passage will illustrate the enormous difficulty which the notion of the "moving force" emanating from the sun presents to Kepler's mind.

"Though the light of the sun cannot itself be the moving force . . . it may perhaps represent a kind of vehicle, or tool, which the moving force uses. But the following considerations seem to contradict this. First, the light is arrested in regions that lie in shade. If, then, the moving force were to use light as a vehicle, then darkness would bring the planets to a standstill. . . .

Since there is as much of this force present in the wider, distant orbits as in nearer and narrower ones, it follows that nothing of this force is lost on the journey from its source; nothing is dispersed between the source and the star. This emanation is therefore unsubstantial as light is, and not accompanied by a loss of substance, as are the emanations of odors, or of the heat which goes out from a glowing stove, and the like, where the intervening space is filled [by the emanation]. We must, therefore, conclude that, just as the light which lights up everything on earth is a non-substantial variety of the fire in the solar body, likewise this force which grips and carries the planet bodies is a non-substantial variety of the force which has its seat in the sun itself; and that it has immeasurable strength, and thus gives the first impulse to all motion in the world. . . .

This kind of force, just as the kind of force which is light . . . cannot be regarded as something which expands into the space between its source and the movable body, but as something which the movable body receives out of the space which it occupies. . . .* It is propagated through the universe . . . but it is nowhere received except where there is a movable body, such as a planet. The answer to this is: although the moving force has no substance, it is aimed at substance, i.e., at the planet body to be moved. . . .

Who, I ask, will pretend that light has substance? Yet nevertheless it acts and is acted upon in space, it is refracted and reflected, and it has quantity, so that it may be dense or sparse, and can be regarded as a plane where it is received by something capable of being lit up. For, as I said in my *Optics*, the same thing applies to light as to our moving force: it has no present existence in the space between the source and the object which it lights up, although it has passed through that space in the past; it 'is' not, it 'was,' so to speak."[40]

The contemporary physicist grappling with the paradoxes of relativity and quantum mechanics will find here an echo of his perplexities. In the end, Kepler managed to get to terms with his "moving force" by visualizing it as a vortex, "a raging current which tears all the planets, and perhaps all the celestial ether, from West to East."[41] But he was nevertheless compelled to ascribe to each planet a kind of mind that enables it to recognize its position in space, and to adjust its reactions accordingly. To careless readers of the *Astronomia Nova* this looked as if the animal spirits had gained readmission into the model which he intended to be a purely mechanical clockwork—like ghosts who cannot resign themselves to their final banishment from the world of the living. But Kepler's planetary minds bear in fact no resemblance to those medieval planet-moving angels and spirits. They

* Note that this description is closer to the modern notion of the gravitational or electro-magnetic *field* than to the classic Newtonian concept of *force*.

have no "souls," only "minds"; no sense organs, and no will of their own; they are rather like the computing machines in guided missiles.

"O Kepler, dost thou wish then to equip each planet with two eyes? Not at all. For it is not necessary, either, to attribute them feet or wings to enable them to move. . . . Our speculations have not yet exhausted all Nature's treasures, to enable us to know how many senses exist. . . .

The subtle reflections of some people concerning the blessed angels' and spirits' nature, motions, places, and activities, do not concern us here. We are discussing natural matters of much lower rank: forces which do not exercise free will when they change their activities, intelligences which are by no means separate from, but attached to, the stellar bodies to be moved, and are one with them."[42]

Thus the function of the planet's mind is confined to responding in a lawful, orderly, and therefore "intelligent" manner to the various forces tugging at it. It is really a superior kind of electronic brain with an Aristotelian bias. Kepler's ambiguity is, in the last analysis, merely a reflection of the mind-matter dilemma, which becomes particularly acute in periods of transition—including our own. As his most outstanding German biographer has put it:

"The physical expositions of Kepler have a special message to those who feel the need to inquire into the first beginnings of the mechanistic explanation of nature. He touches, indeed, on the profoundest questions of the philosophy of nature when he confronts, in his subtle manner, the concepts of *mens* and *natura*, compares their pragmatic values, and delimits their fields of application. Have we outgrown this antithesis in our day? Only those will believe that who are unaware of the metaphysical nature of our concept of physical force. . . . At any rate, Kepler's explanations may serve as a stimulus to a wholesome contemplation of the axioms and limits of mechanistic philosophy in our time of widespread and disastrous scientific dogmatism."[43]

Though Kepler was unable to solve the dilemma, he polished its horns, as it were. The angels, spirits, and unmoved movers were banished from cosmology; he sublimated and distilled the problem to a point where only the ultimate mystery remains. Though he was always attracted, with a mixture of disgust and fascination, by theological disputes, he uncompromisingly and indeed vehemently rejected the incursion of the theologians into science. On this point he made his position very clear in a statement—or rather a battle cry—in the introduction to the *New Astronomy*.

"So much for the authority of Holy Scripture. Now as regards the opinions of the saints about these matters of nature, I answer in one word, that in theology the weight of Authority, but in philosophy the weight of Reason alone is valid. Therefore a saint was Lanctantius, who denied the earth's rotundity; a saint was Augustine, who admitted the rotundity, but denied that antipodes exist. Sacred is the Holy Office of our day, which admits the smallness of the earth but denies its motion: but to me more sacred than all these is Truth, when I, with all respect for the doctors of the Church, demonstrate from philosophy that the earth is round, circumhabited by antipodes, of a most insignificant smallness, and a swift wanderer among the stars."

chapter seven

Kepler Depressed

1. PUBLISHING DIFFICULTIES

The writing of the *New Astronomy* had been an obstacle race over six years.

At the start there had been the quarrels with Tycho, the long sojourns in Gratz, illness, and the drudgery of the pamphlets against Ursus and Craig. When the great Dane died, and Kepler was appointed his successor, he may have hoped to be able to work in peace; instead, his life became even more disorganized. His official and unofficial duties included the publication of annual calendars with astrological predictions; the casting of horoscopes for distinguished visitors at court; the publication of comments on eclipses, comets, and a new star; answering, at great length, the queries on every subject under the sun put to him by the various patrons with whom he corresponded; and above all petitioning, lobbying, and intriguing to obtain at least a fraction of the salaries and printing costs due to him. He had discovered his Second Law as early as 1602, one year after Tycho's death; but the next year he was almost entirely occupied with other labors, among them the great work on optics, published in 1604; the year after that, he became stuck on the egg-shaped orbit, fell ill, and again thought that he was dying; and only around Easter in 1605 was the *New Astronomy* completed in outline.

But it took another four years to get it published. The reason for this delay was lack of money to pay for the printing, and a harassing feud with Tycho's heirs, led by the swashbuckling *Junker* Tengnagel. This character, it will be remembered, had married Tycho's daughter Elisabeth—the only achievement on which he could base his claim to the Tychonic heritage. He was determined to cash in on it, and sold Tycho's observations and instruments to the Emperor for the sum of twenty thousand thalers. But the imperial treasury never paid the *Junker*; he had to content himself with an annual five per cent interest on the debt—which was still twice the amount of Kepler's salary. As a result, Tycho's instruments, the wonder of the world, were kept by Tengnagel behind lock and key; within a few years they decayed to scrap metal. A similar fate would no doubt have befallen the treasure of Tycho's observations, if Kepler had not hurriedly pinched them for the benefit of posterity. In a letter to one of his English admirers[1] he calmly reported:

"I confess that when Tycho died, I quickly took advantage of the absence, or lack of circumspection, of the heirs, by taking the observations under my care, or perhaps usurping them. . . ."

It had always been his avowed intention to get possession of Tycho's treasure, and he had succeeded.

The Tychonides were understandably furious; Kepler, the introspective grave robber, quite saw their point.

"The cause of this quarrel lies in the suspicious nature and bad manners of the Brahe family, but on the other hand also in my own passionate and mocking character. It must be admitted that Tengnagel had important reasons for suspecting me. I was in possession of the observations and refused to hand them over to the heirs. . . ."[2]

The negotiations dragged on for several years. The *Junker*, ambitious, stupid, and vain, proposed a dirty deal: he would keep his peace if all Kepler's future works were published under their joint names. Astonishingly, Kepler agreed: he was always strangely indifferent to the fate of his published works. But he asked that the *Junker* should, in exchange, hand over a quarter of the annual thousand thalers which he drew from the treasury. This Tengnagel refused, considering two hundred and fifty a year to be too high a price for immortality. He thus deprived future scholars of a delightful subject of controversy on the question: Which of the two partners discovered the Tengnagel-Kepler laws.

In the meantime, the *Junker* had embraced the Catholic faith and been made an appellate councilor at court. This enabled him to impose his conditions on Kepler, and made it impossible for Kepler to publish his book without Tengnagel's consent. Thus Kepler found himself "tied hands and feet," while the *Junker* "sits like a dog in the manger, unable to put the treasure to use, and preventing others from doing so."[2a] A compromise was reached at last: Tengnagel gave his gracious consent to the printing of the *New Astronomy* on condition that it should carry a preface from his own pen. Its text is printed in Note 3. If Osiander's preface to the *Book of the Revolutions* displayed the wisdom of a gentle snake, in Tengnagel's preface to the *New Astronomy* we hear the braying of a pompous ass echoing down the centuries.

At last, in 1608, the printing of the book could begin; it was finished in the summer of 1609, in Heidelberg, under Kepler's supervision. It was a beautifully printed volume in folio, of which only a few copies survive. The Emperor claimed the whole edition as his property and forbade Kepler to sell or give away any copy of it "without our foreknowledge and consent." But since his salary was in arrears, Kepler felt at liberty to do as he liked, and sold the whole edition to the printers. Thus the story of the *New Astronomy* begins and ends with acts of larceny, committed *ad majorem Dei gloriam*.

2. RECEPTION OF THE "ASTRONOMIA NOVA"

How far ahead of his time Kepler was—not merely by his discoveries, but in his whole manner of thought—one can gather from the negative reactions of his friends and correspondents. He received no help, no encouragement; he had patrons and well-wishers, but no congenial spirit.

Old Maestlin had been silent for the last five years, in spite of a persistent stream of letters from Kepler, who kept his old teacher informed of every important event in his life and researches. Just before the completion of the *New Astronomy*, Maestlin broke his silence with a very moving letter which, however, was a complete letdown insofar as Kepler's hopes for guidance, or at least of shared interests, were concerned.

"Tuebingen, January 28, 1605.

Although I have for several years neglected writing to you, your steadfast attachment, gratitude, and sincere affection have not weakened but become rather stronger, albeit you have reached such a high step and distinguished position that you could, if you wished, look down on me. . . . I do not wish to apologize further, and say only this: I have nothing of the same value to offer in writing to such an outstanding mathematician. . . . I must further confess that your questions were sometimes too subtle for my knowledge and gifts, which are not of the same stature. Hence I could only keep silent. . . . You will wait in vain for my criticism of your book on optics, which you request so urgently; it contains matters too lofty for me to permit myself to judge it. . . . I congratulate you.

163

The frequent and most flattering mention of my name [in that book] is a special proof of your attachment. But I fear that you credit me with too much. If only I were such as your praise makes me appear. But I understand only my modest craft."[4]

And that was the end of it, though Kepler persisted in the one-sided correspondence, and also in his miscellaneous requests—Maestlin should make inquiries about the suitor of Kepler's sister, Maestlin should find him an assistant, and so forth—which the old man steadfastly ignored.

The most detailed letters about the progress of the *New Astronomy* Kepler wrote to David Fabricius, a clergyman and amateur astronomer in Friesland. Some of these letters cover over twenty, and up to forty, foolscap pages. Yet he could never persuade Fabricius to accept the Copernican view; and when Kepler informed him of his discovery of the First Law, Fabricius' reaction was:

"With your ellipse you abolish the circularity and uniformity of the motions, which appears to me the more absurd the more profoundly I think about it. . . . If you could only preserve the perfect circular orbit, and justify your elliptic orbit by another little epicycle, it would be much better."[5]

As for the patrons and well-wishers, they tried to encourage him, but were unable to comprehend what he was up to. The most enlightened among them, the physician Johannes Brengger, whose opinion Kepler particularly valued, wrote:

"When you say that you aim at teaching both a new physics of the sky and a new kind of mathematics, based not on circles but on magnetic and intelligent forces, I rejoice with you, although I must frankly confess that I am unable to imagine, and even less to comprehend, such a mathematical procedure."[6]

This was the general reaction of Kepler's contemporaries in Germany. It was summed up by one of them:

"In trying to prove the Copernican hypothesis from physical causes, Kepler introduces strange speculations which belong not in the domain of astronomy, but of physics."[7]

Yet a few years later the same man confessed:

"I no longer reject the elliptical form of the planetary orbits and allowed myself to be persuaded by the proofs in Kepler's work on Mars."[8]

The first to realize the significance and implications of Kepler's discoveries were neither his German compatriots nor Galileo in Italy, but the British: the traveler Edmund Bruce, the mathematician Thomas Harriot, tutor of Sir Walter Raleigh; the Reverend John Donne, the astronomical genius Jeremiah Horrocks, who died at twenty-one; and, last, Newton.

3. ANTICLIMAX

Delivered from his monumental labors, Kepler sank into the usual anticlimax.

He turned back to his persistent dream, the harmony of the spheres, convinced that the whole *New Astronomy* was merely a steppingstone toward that ultimate aim in his "sweating and panting pursuit of the Creator's tracks."[9] He published two polemical works on astrology, a pamphlet on comets, another about the shape of snow crystals, conducted a voluminous correspondence on the true date of the birth of Christ. He continued with his calendars and weather predictions: on one occasion, when a violent thunderstorm darkened the sky at noon, as he had predicted a fortnight earlier, the people in the streets of Prague yelled, pointing at the clouds, "It's that Kepler coming."

He was by now an internationally famous scholar, a member of the Italian Accademia dei Lincei (a forerunner of the Royal Society), but even more pleased about the distinguished society in which he moved in Prague.

"The imperial councilor and first secretary, Johann Polz, is very fond of me. [His wife and] the whole family are conspicuous here in Prague for their Austrian elegance and their distinguished and noble manners; so that it would be due to their influence if on some future day I made some progress in this respect, though, of course, I am still far away from it. . . . Notwithstanding the shabbiness of my household and my low rank (for they are considered to belong to the nobility), I am free to come and go in their house as I please."[10]

His rise in social status was reflected in the personalities of the godparents to the two children who were born to him in Prague: the wives of halberdiers to the first; counts of the palatinate and ambassadors to the second. There was an endearing Chaplinesque quality about Kepler's efforts to display social graces. "What a job, what an upheaval to invite fifteen to sixteen women to visit my wife in childbed, to play host to them, to compliment them to the door!"[10a] Though he wore fine cloth and Spanish ruffles, his salary was always in arrears. "My hungry stomach looks up like a little dog to its master who used to feed it."[11]

Visitors to Prague were invariably impressed by his dynamic personality and quicksilvery mind; yet he was still suffering from lack of self-assurance—a chronic ill, on which success acted as a temporary sedative, but never as a complete cure. The turbulent times increased his feeling of insecurity; he lived in constant fear of penury and starvation, complicated by his obsessive hypochondria.

"You inquire after my illness? It was an insidious fever which originated in the gall and returned four times because I repeatedly sinned in my diet. On May 29 my wife forced me, by her pesterings, to wash, for once, my whole body. She immersed me in a tub (for she has a horror of public baths) with well-heated water; its heat afflicted me and constricted my bowels. On May 31 I took a light laxative, according to habit. On June 1, I bled myself, also according to habit: no urgent disease, not even the suspicion of one, compelled me to do it, nor any astrological consideration. . . . After losing blood, I felt well for a few hours; but in the evening an evil sleep threw me on my mattress and constricted my guts. Sure enough, the gall at once gained access to my head, bypassing the bowels. . . . I think I am one of those people whose gall bladder has a direct opening into the stomach; such people are short-lived as a rule."[12]

Even without hypochondria, there were sufficient reasons for anxiety. His imperial patron sat on a quaking throne—though, in truth, Rudolph rarely sat on it, preferring to hide from his odious fellow creatures among his mechanical clocks and toys, gems and coins, retorts and alembics. There were wars and rebellions in Moravia and Hungary, and the treasury was empty. As Rudolph progressed from eccentricity to apathy and melancholia, his brother was depriving him piecemeal of his domains; in a word, Rudolph's final abdication was only a question of time. Poor Kepler, already expelled from his livelihood in Gratz, saw a second exile looming before him, and had to start once more pulling wires, stretching out feelers, and clutching at straws. But the

Lutheran worthies in his beloved Wuerttemberg would
have nothing to do with their *enfant terrible*, and Maxi-
milian of Bavaria turned a politely deaf ear, as did other
princes whom he approached. The year after the publication of the *New Astronomy* saw him at his lowest ebb,
unable to do any serious work, "my mind prostrate in a
pitiful frost."

Then came an event which not only thawed it, but set
it to bubble and boil.

4. THE GREAT NEWS

One day in March 1610, a certain Herr Johannes Mat-
thaeus Wackher von Wackenfels, Privy Councilor to His
Imperial Majesty, Knight of the Golden Chain and of
the Order of St. Peter, amateur philosopher and poet,
drove up in his coach to Kepler's house and called for
him in great agitation. When Kepler came down, Herr
Wackher told him the news had just arrived at court
that a mathematician named Galileus in Padua had
turned a Dutch spyglass at the sky and discovered
through its lenses four new planets in addition to the
five which had always been known.

> "I experienced a wonderful emotion while I lis-
> tened to this curious tale. I felt moved in my deep-
> est being. . . . [Wackher] was full of joy and
> feverish excitement; at one moment we both
> laughed at our confusion, the next he continued his
> narrative and I listened intently—there was no end
> to it. . . ."[13]

Wackher von Wackenfels was twenty years older than
Kepler, and devoted to him. Kepler sponged on the privy
councilor's excellent wine, and had dedicated to him his
treatise on the snow crystals as a New Year's gift. Wackher, though a Catholic convert, believed in the plurality
of worlds; accordingly, he thought that Galileo's discov-

eries were planets to other stars, outside our solar system. Kepler rejected this idea; but he equally refused to admit that the new heavenly bodies could be revolving around the sun, on the grounds that since there were only five perfect solids, there could only be six planets —as he had proved to his own satisfaction in the *Cosmic Mystery*. He accordingly deduced a priori that what Galileo had seen in the sky could only be secondary satellites, which circled around Venus, Mars, Jupiter, and Saturn, as the moon circled around the earth. Once again he had guessed nearly right for the wrong reasons: Galileo's discoveries were indeed moons, but all four of them were moons of Jupiter.

A few days later, authentic news arrived in the shape of Galileo's short but momentous booklet, *Sidereus Nuncius*, the *Messenger from the Stars*.[14] It heralded the assault on the universe with a new weapon, an optic battering ram, the telescope.

chapter eight
Kepler and Galileo

1. A DIGRESSION ON MYTHOGRAPHY

It was indeed a new departure. The range and power of the main sense organ of Homo sapiens had suddenly started to grow in leaps to thirty times, a hundred times, a thousand times its natural capacity. Parallel leaps and bounds in the range of other organs were soon to transform the species into a race of giants in power—without enlarging his moral stature by an inch. It was a monstrously one-sided mutation—as if moles were growing to the size of whales, but retaining the instincts of moles. The makers of the scientific revolution were individuals who in this transformation of the race played the part of the mutating genes. Such genes are *ipso facto* unbalanced and unstable. The personalities of these "mutants" already foreshadowed the discrepancy in the next development of man: the intellectual giants of the scientific revolution were moral dwarfs.

They were, of course, neither better nor worse than the average of their contemporaries. They were moral dwarfs only in proportion to their intellectual greatness. It may be thought unfair to judge a man's character by the standard of his intellectual achievements, but the great civilizations of the past did precisely this; the divorce of moral from intellectual values is itself a characteristic development of the last few centuries. It is foreshadowed in the philosophy of Galileo, and became fully explicit in the ethical neutrality of modern determinism. The indulgence with which historians of science treat the founding fathers is based on precisely that tradition which the fathers introduced—the tradition of keeping intellect and character as strictly apart as Galileo taught us to separate the "primary" and "secondary" qualities of objects. Thus moral assessments are thought to be essential in the case of Cromwell or Danton, but irrelevant in the case of Galileo, Descartes, or Newton. However, the scientific revolution produced not only discoveries, but a new attitude to life, a change in the philosophical climate. And on that new climate, the personalities and beliefs of those who initiated it had a lasting influence. The most pronounced of these influences, in their different fields, were Galileo's and Descartes'.

The personality of Galileo, as it emerges from works of popular science, has little relation to historic fact. This is not caused by a benevolent indifference toward the individual as distinct from his achievement, but by more partisan motives. In works with a theological bias, he appears as the nigger in the woodpile; in rationalist mythography, as the Maid of Orleans of science, the St. George who slew the dragon of the Inquisition. It is, therefore, hardly surprising that the fame of this outstanding genius rests mostly on discoveries he never made, and on feats he never performed. Contrary to statements in even recent outlines of science, Galileo did not invent the telescope; nor the microscope; nor the thermometer; nor the pendulum clock. He did not discover the law of inertia; nor the parallelogram of forces or motions; nor the sunspots. He made no contribution to theoretical astronomy; he did not throw down weights from the leaning tower of Pisa, and did not prove the truth of the Copernican system. He was not tortured by the Inquisition, did not languish in its dungeons, did not say *"eppur si muove"*; and he was not a martyr of science.

What he *did* was to found the modern science of dynamics, which makes him rank among the men who shaped human destiny. It provided the indispensable complement to Kepler's laws for Newton's universe. "If I have been able to see farther," Newton said, "it was because I stood on the shoulders of giants." The giants were, chiefly, Kepler, Galileo, and Descartes.

2. YOUTH OF GALILEO

Galileo Galilei was born in 1564 and died in 1642, the year Newton was born. His father, Vincento Galilei, was an impoverished scion of the lower nobility, a man of remarkable culture, with considerable achievements as a composer and writer on music, a contempt for authority, and radical leanings. He wrote, for instance (in a study on counterpoint), "It appears to me that those who try to prove an assertion by relying simply on the weight of authority act very absurdly."[1]

One feels at once the contrast in climate between the childhoods of Galileo and those of Copernicus, Tycho, and Kepler, who never completely severed the navel cord which had fed into them the rich, mystic sap of the Middle Ages. Galileo is a second-generation intellectual, a second-generation rebel against authority; in a nineteenth-century setting, he would have been the Socialist son of a Liberal father.

His early portraits show a ginger-haired, short-necked, beefy young man of rather coarse features, a thick nose, and conceited stare. He went to the excellent Jesuit school at the monastery of Vallombrosa, near Florence; but Galileo Senior wanted him to become a merchant (which was by no means considered degrading for a patrician in Tuscany) and brought the boy home to Pisa; then, in recognition of his obvious gifts, changed his mind and at seventeen sent him to the local university to study medicine. But Vincento had five children to look after (a younger son, Michelangelo, plus three daughters), and the university fees were high; so he tried to obtain a scholarship for Galileo. Although there were no less than forty scholarships for poor students available in Pisa, Galileo failed to obtain one, and was compelled to leave the university without a degree. This is the more surprising as he had already given unmistakable proof of his brilliance: in 1582, in his second year at the university, he discovered the fact that a pendulum of a given length swings at a constant frequency, regardless of amplitude.[2] His invention of the "pulsilogium," a kind of metronome for timing the pulse of patients, was probably made at the same time. In view of this and other proofs of the young student's mechanical genius, his early biographers explained the refusal of a scholarship by pointing to the animosity which his unorthodox anti-Aristotelian views raised. In fact, however, Galileo's early views on physics contain nothing of a revolutionary nature.[3] It is more likely that the refusal of the scholarship was due, not to the unpopularity of Galileo's views, but to his person—that cold, sarcastic presumption by which he managed to spoil his case throughout his life.

Back home he continued his studies, mostly in applied mechanics, which attracted him more and more, perfecting his dexterity in making mechanical instruments and gadgets. He invented a hydrostatic balance, wrote a treatise on it which he circulated in manuscript, and began to attract the attention of scholars. Among these was the Marchese Guidobaldo del Monte, who recommended Galileo to his brother-in-law, Cardinal del Monte, who in turn recommended him to Ferdinand de Medici, the ruling duke of Tuscany; as a result, Galileo was appointed a lecturer in mathematics at the University of Pisa, four years after that same university had refused him a scholarship. Thus, at the age of twenty-five, he was launched on his academic career. Three years later, in 1592, he was appointed to the vacant Chair of Mathematics at the famous University of Padua, again through the intervention of his patron, del Monte.

Galileo remained in Padua for eighteen years, the most creative years of his life. It was here that he laid the foundations of modern dynamics, the science concerned with moving bodies. But the results of these researches he published only toward the end of his life. Up to the age of forty-six, when the *Messenger from the Stars* was sent into the world, Galileo had published no scientific work.[4] His growing reputation in this period, before his discoveries through the telescope, rested partly on treatises and lectures circulated in manuscript, partly on his mechanical inventions (among them the thermoscope, a forerunner of the thermometer), and the instruments which he manufactured in large numbers with skilled artisans in his own workshop. But his truly great discoveries—such as the laws of motion of falling bodies and projectiles—and his ideas on cosmology he kept strictly for himself and for his private correspondents. Among these was Johannes Kepler.

3. THE CHURCH AND THE COPERNICAN SYSTEM

The first contact between the two founding fathers took place in 1597. Kepler was then twenty-six, a professor of mathematics in Gratz; Galileo was thirty-three, a professor of mathematics in Padua. Kepler had just completed his *Cosmic Mystery* and, profiting from a friend's journey to Italy, had sent copies of it, among others, "to a mathematician named Galileus Galileus, as he signs himself."[5]

Galileo acknowledged the gift in the following letter.

"Your book, my learned doctor, which you sent me through Paulus Amberger, I received not a few days but merely a few hours ago; since the same Paulus informed me of his impending return to Germany, I would be ungrateful indeed not to thank you at once: I accept your book the more gratefully as I regard it as proof of having been found worthy of your friendship. So far I have only perused the preface of your work, but from this I gained some notion of its intent,* and I indeed congratulate myself on having an associate in the study of Truth who is a friend of Truth. For it is a misery that so few exist who pursue the Truth and do not pervert philosophical reason. However, this is not the place to deplore the miseries of our century but to congratulate you on the ingenious arguments you found in proof of the Truth. I will only add that I promise to read your book in tranquillity, certain to find the most admirable things in it, and this I shall do the more gladly as I adopted the teaching of Copernicus many years ago, and his point of view enables me to explain many phenomena of nature which certainly remain inexplicable according to the more current hypotheses. I have written [*conscripsi*] many arguments in support of him and in refutation of the opposite view —which, however, so far I have not dared to bring into the public light, frightened by the fate of Copernicus himself, our teacher, who, though he acquired immortal fame with some, is yet to an infinite multitude of others (for such is the number of fools) an object of ridicule and derision. I would certainly dare to publish my reflections at once if more people like you existed; as they don't, I shall refrain from doing so."

* The preface (and first chapter) proclaim Kepler's belief in the Copernican system and outline his arguments in favor of it.

There follow more polite affirmations of esteem, the signature "Galileus Galileus," and the date: August 4, 1597.[6]

The letter is important for several reasons. First, it provides conclusive evidence that Galileo had become a convinced Copernican in his early years. He was thirty-three when he wrote the letter; and the phrase "many years ago" indicates that his conversion took place in his twenties. Yet his first explicit public pronouncement in favor of the Copernican system was only made in 1613, a full sixteen years after his letter to Kepler, when Galileo was forty-nine years of age. Through all these years he not only taught, in his lectures, the old astronomy according to Ptolemy, but expressly repudiated Copernicus. In a treatise which he wrote for circulation among pupils and friends, of which a manuscript copy, dated 1606, survives,[6a] he adduced all the traditional arguments against the earth's motion: that rotation would make it disintegrate, that clouds would be left behind, etc., etc.—arguments which, if the letter is to be believed, he himself had refuted many years before.

But the letter is also interesting for other reasons. In a single breath, Galileo four times evokes "Truth": friend of Truth, investigating Truth, pursuit of Truth, proof of Truth; then, apparently without awareness of the paradox, he calmly announces his intention to suppress Truth. This may partly be explained by the mores of late Renaissance Italy ("that age without a superego," as a psychiatrist described it); but taking that into account, one still wonders at the motives of his secrecy.

Why, in contrast to Kepler, was he so afraid of publishing his opinions? He had, at that time, no more reason to fear religious persecution than Copernicus had. The Lutherans, not the Catholics, had been the first to attack the Copernican system—which prevented neither Rheticus nor Kepler from defending it in public. The Catholics, on the other hand, were uncommitted. In Co-

pernicus' own day, they were favorably inclined toward
him. Twenty years after publication of Copernicus'
book, the Council of Trent redefined Church doctrine
and policy in all its aspects, but it had nothing to say
against the heliocentric system of the universe. Galileo
himself, as we shall see, enjoyed the active support of a
galaxy of cardinals, including the future Urban VIII,
and of the leading astronomers among the Jesuits. Up to
the fateful year 1616 discussion of the Copernican system was not only permitted, but encouraged by them
—under the one proviso, that it should be confined to
the language of science, and should not impinge on
theological matters. The situation was summed up
clearly in a letter from Cardinal Dini to Galileo in 1615.
"One may write freely as long as one keeps out of the
sacristy."[7] This was precisely what the disputants failed
to do, and it was at this point that the conflict began.
But nobody could have foreseen these developments
twenty years earlier, when Galileo wrote to Kepler.

Thus legend and hindsight combined to distort the
picture, and gave rise to the erroneous belief that to
defend the Copernican system as a working hypothesis
entailed the risk of ecclesiastical disfavor or persecution.
During the first fifty years of Galileo's lifetime no such
risk existed; and the thought did not even occur to Galileo. What he feared is clearly stated in his letter: to
share the fate of Copernicus, to be mocked and derided;
ridendus et explodendum—"laughed at and hissed off
the stage" are his exact words. Like Copernicus, he was
afraid of the ridicule of both the unlearned and the
learned asses, but particularly of the latter: his fellow
professors at Pisa and Padua, the stuffed shirts of the
Peripatetic school, who still considered Aristotle and
Ptolemy as absolute authority. And this fear, as will be
seen, was fully justified.

4. EARLY QUARRELS

Young Kepler was delighted with Galileo's letter. On the first occasion when a traveler left Gratz for Italy, he answered in his impulsive manner:

"Gratz, October 13, 1597.

Your letter, my most excellent humanist, which you wrote on August 4, I received on September 1; it caused me to rejoice twice: first because it meant the beginning of a friendship with an Italian; secondly, because of our agreement on the Copernican cosmography. . . . I assume that if your time has permitted it, you have by now become better acquainted with my little book, and I ardently desire to know your critical opinion of it; for it is my nature to press all to whom I write for their unvarnished opinion; and believe me, I much prefer even the most acrimonious criticism of a single enlightened man to the unreasoned applause of the common crowd.

I would have wished, however, that you, possessed of such an excellent mind, took up a different position. With your clever secretive manner you underline, by your example, the warning that one should retreat before the ignorance of the world, and should not lightly provoke the fury of the ignorant professors; in this respect you follow Plato and Pythagoras, our true teachers. But considering that in our era, at first Copernicus himself and after him a multitude of learned mathematicians have set this immense enterprise going so that the motion of the earth is no longer a novelty, it would be preferable that we help to push home by our common efforts this already moving carriage to its destination. . . . You could help your comrades, who labor under such iniquitous criticism, by giving them the comfort of your agreement and the protection of your authority. For not only your Italians refuse to believe that they are in motion because they do not feel it; here in Germany, too, one does not make oneself popular by holding such opinions. But there exist arguments which protect us in the face of these difficulties. . . . Have faith, Galilii, and come forward! If my guess is right, there are but few among the prominent mathematicians of Europe who would wish to secede from us: for such is the force of Truth. If your Italy seems less advantageous to you for publishing [your works] and if your living there is an obstacle, perhaps our Germany will allow us to do so. But enough of this. Let me know, at least privately if you do not want to do it in public, what you have discovered in support of Copernicus. . . ."

Kepler then confessed that he had no instruments, and asked Galileo whether he had a quadrant sufficiently precise to read quarter-minutes of arc; if so, would Galileo please make a series of observations to prove that the fixed stars show small seasonal displacements—which would provide direct proof of the earth's motion.

"Even if we could detect no displacement at all, we would nevertheless share the laurels of having investigated a most noble problem which nobody has attacked before us. *Sat Sapienti*. . . . Farewell, and answer me with a very long letter."[8]

Poor naïve Kepler! It did not occur to him that Galileo might take offense at his exhortations, and regard them as an implied reproach of cowardice. He waited in vain for an answer to his exuberant overtures. Galileo withdrew his feelers; for the next twelve years, Kepler did not hear from him.

But from time to time unpleasant rumors reached him from Italy. Among Kepler's admirers was a certain Edmund Bruce, a sentimental English traveler in Italy, amateur philosopher and science snob, who loved to rub shoulders with scholars and to spread gossip about them. In August 1602, five years after Galileo had broken off their correspondence, Bruce wrote Kepler from Florence that Magini (the professor of astronomy at Bologna) had assured him of his love and admiration of Kepler, whereas Galileo had admitted to him, Bruce, having received Kepler's *Mysterium*, but had denied this to Magini.

"I scolded Galileo for his scant praise of you, for I know for certain that he lectures on your and his own discoveries to his pupils and others. I, however, act and shall always act in a manner which serves not his fame, but yours."[9]

Kepler could not be bothered to answer this busybody, but a year later—August 21, 1603—Bruce wrote again, this time from Padua.

"If you knew how often and how much I discuss you with all the savants of Italy you would consider me not only an admirer but a friend. I spoke with them of your admirable discoveries in music, of your studies of Mars, and explained to them your *Mysterium* which they all praise. They wait impatiently for your future works. . . . Galileo has your book and teaches your discoveries as his own. . . ."[10]

This time Kepler did answer. After apologizing for the delay and declaring himself delighted with Bruce's friendship, he continued:

"But there is something about which I wish to warn you. Do not form a higher opinion of me, and do not induce others to do so, than my achievements are able to justify. . . . For you certainly understand that betrayed expectations lead eventually

to contempt. I wish in no way to restrain Galileo
from claiming what is mine as his own. My witnesses are the bright daylight and time."[11]

The letter ends with "Greetings to Magini and
Galileo."

Bruce's accusations should not be taken seriously. In
fact, the opposite is true. the trouble with Galileo was
not that he appropriated Kepler's discoveries—but that
he ignored them, as we shall see. But the episode nevertheless sheds some additional light on the relations between the two men. Though Bruce cannot be trusted
on points of fact, the inimical attitude of Galileo to
Kepler emerges clearly from Bruce's letters. It fits in
with the fact that he broke off the correspondence, and
with later events.

Kepler, on the other hand, who had good reason to
be offended by Galileo's silence, could easily have been
provoked by Bruce's scandalmongering into starting one
of those juicy quarrels between scholars which were the
order of the day. He was suspicious and excitable
enough, as his relations with Tycho have shown. But
toward Galileo he always behaved in an oddly generous
way. It is true that they lived in different countries and
never met personally; but hatred, like gravity, is capable
of action at a distance. The reason for Kepler's forbear-
ance was perhaps that he had no occasion to develop
an inferiority complex toward Galileo.

The year after the Bruce episode, in October 1604, a
bright new star appeared in the constellation Serpen-
tarius. It caused even more excitement than Tycho's
famous nova of 1572, because its appearance happened
to coincide with a so-called great conjunction of Jupiter, Saturn, and Mars in the "fiery triangle"—a gala performance that occurs only once in every eight hundred
years. Kepler's book *De Stella Nova* (1606) was prima-
rily concerned with its astrological significance; but he

showed that the nova, like the previous one, must be located in the "immutable" region of the fixed stars, and thus drove another nail into the coffin of the Aristotelian universe. The star of 1604 is still called "Kepler's nova."*

Galileo, too, observed the new star, but published nothing about it. He gave three lectures on the subject, of which only fragments are preserved; he too seems to have denied the contention of the Aristotelians that it was a meteor or some other sublunary phenomenon, but could not have gone much further, since his lectures in defense of Ptolemy were still circulated two years later.[12]

Between 1600 and 1610 Kepler published his *Optics* (1604), the *New Astronomy* (1609), and a number of minor works. In the same period, Galileo worked on his fundamental researches into free fall, the motion of projectiles, and the laws of the pendulum, but published nothing except a brochure containing instructions for the use of the so-called military or proportional compass. This was an invention made in Germany some fifty years earlier,[13] which Galileo had improved, as he improved a number of other gadgets that had been known for a long time. Out of this minor publication[14] developed the first of the futile and pernicious feuds which Galileo was to wage all his life.

It began when a mathematician named Balthasar Capra in Padua published, a year after Galileo, another brochure of instructions for the use of the proportional compass.[15] Galileo's *Instructions* were in Italian,

---

* John Donne referred to Kepler's nova when he wrote (*To the Countesse of Huntingdon*):

> Who vagrant transitory Comets sees,
> Wonders, because they are rare: but a new starre
> Whose motion with the firmament agrees,
> Is miracle, for there no new things are.

Capra's in Latin; both referred to the same subject, which interested only military engineers and technicians. It is very likely that Capra had borrowed from Galileo's *Instructions* without naming him; on the other hand, Capra showed that some of Galileo's explanations were mathematically erroneous, but again without naming him. Galileo's fury knew no bounds. He published a pamphlet *Against the Calumnies and Impostures of Balthasar Capra, etc.* (Venice, 1607), in which that unfortunate man and his teacher[16] were described as "that malevolent enemy of honor and of the whole of mankind," "a venomspitting basilisk," "an educator who bred the young fruit on his poisoned soul with stinking ordure," "a greedy vulture, swooping at the unborn young to tear its tender limbs to pieces," and so on. He also obtained from the Venetian court the confiscation, on the grounds of plagiarism, of Capra's *Instructions*. Not even Tycho and Ursus had sunk to such fishwife language; yet they had fought for the authorship of a system of the universe, not of a gadget for military engineers.

In his later polemical writings, Galileo's style progressed from coarse invective to satire, which was sometimes cheap, often subtle, always effective. He changed from the cudgel to the rapier, and achieved a rare mastery of it, while in the purely expository passages his lucidity earned him a prominent place in the development of Italian didactic prose. But behind the polished façade, the same passions were at work which had exploded in the affair of the proportional compass: vanity, jealousy, and self-righteousness combined into a demoniac force, which drove him to the brink of self-destruction. He was utterly devoid of any mystical, contemplative leanings, in which the bitter passions could from time to time be resolved; he was unable to transcend himself and find refuge, as Kepler did in his darkest hours, in the cosmic mystery. He did not stand astride the watershed; Galileo is wholly and frighteningly modern.

5. THE IMPACT OF THE TELESCOPE

It was the invention of the telescope that brought Kepler and Galileo, each traveling along his own orbit, to their closest conjunction. To pursue the metaphor, Kepler's orbit reminds one of the parabola of comets which appear from infinity and recede into it; Galileo's as an eccentric ellipse, closed upon itself.

The telescope was, as already mentioned, not invented by Galileo. In September 1608 a man at the annual Frankfurt fair offered a telescope for sale which had a convex and a concave lens, and magnified seven times. On October 2, 1608, the spectacles maker Johann Lippershey of Middleburg claimed a license for thirty years from the Estates General of the Netherlands for manufacturing telescopes with single and double lenses. In the following month, he sold several of these, for three hundred and six hundred gilders respectively, but was not granted an exclusive license because in the meantime two other men had claimed the same invention. Two of Lippershey's instruments were sent as a gift by the Dutch government to the king of France; and in April 1609 telescopes could be bought in spectacles makers' shops in Paris. In the summer of 1609 Thomas Harriot in England made telescopic observations of the moon, and drew maps of the lunar surface. In the same year several of the Dutch telescopes found their way to Italy and were copied there.

Galileo himself claimed in the *Messenger from the Stars* that he had merely read reports of the Dutch invention, and that these had stimulated him to construct an instrument on the same principle, which he succeeded in doing, "through deep study of the theory of refraction." Whether he actually saw and handled one of the Dutch instruments brought to Italy is a question without importance, for once the principle was known, lesser minds than Galileo's could and did construct similar gadgets. On August 8, 1609, he invited the Venetian Senate to examine his spyglass from the tower of St. Marco, with spectacular success; three days later, he made a present of it to the Senate, accompanied by a letter in which he explained that the instrument, which magnified objects nine times, would prove of utmost importance in war. It made it possible to see "sails and shipping that were so far off that it was two hours before they were seen with the naked eye, steering full sail into the harbor,"[17] thus being invaluable against invasion by sea. It was not the first and not the last time that pure research, that starved cur, snapped up a bone from the warlords' banquet.

The grateful Senate of Venice promptly doubled Galileo's salary to a thousand scudi per year, and made his professorship at Padua (which belonged to the Republic of Venice) a lifelong one. It did not take the local spectacles makers long to produce telescopes of the same magnifying power, and to sell in the streets for a few scudi an article which Galileo had sold the Senate for a thousand a year—to the great amusement of all good Venetians. Galileo must have felt his reputation threatened, as in the affair of the military compass; but, fortunately, this time his passion was diverted into more creative channels. He began feverishly to improve his telescope, and to aim it at the moon and stars, which previously had attracted him but little. Within the next eight months he succeeded, in his own words: "by sparing neither labor nor expense, in constructing for myself an instrument so superior that objects seen through it appear magnified nearly a thousand times, and more than thirty times nearer than if viewed by the natural powers of sight alone."

The quotation is from *Sidereus Nuncius*, the *Messen-*

ger from the Stars, published in Venice in March 1610. It was Galileo's first scientific publication, and it threw his telescopic discoveries like a bomb into the arena of the learned world. It not only contained news of heavenly bodies "which no mortal had seen before"; it was also written in a new, tersely factual style which no scholar had employed before. So new was this language that the sophisticated imperial ambassador in Venice described the *Star Messenger* as "a dry discourse or an inflated boast, devoid of all philosophy."[18] In contrast to Kepler's exuberant baroque style, some passages of the *Sidereus Nuncius* would almost qualify for the austere pages of a contemporary "Journal of Physics."

The whole booklet has only twenty-four leaves in octavo. After the introductory passages, Galileo described his observations of the moon, which led him to conclude:

"that the surface of the moon is not perfectly smooth, free from inequalities, and exactly spherical, as a large school of philosophers considers with regard to the moon and the other heavenly bodies, but that, on the contrary, it is full of irregularities, uneven, full of hollows and protuberances, just like the surface of the earth itself, which is varied everywhere by lofty mountains and deep valleys."

He then turned to the fixed stars, and described how the telescope added, to the moderate numbers that can be seen by the naked eye, "other stars, in myriads, which have never been seen before, and which surpass the old, previously known stars in number more than ten times." Thus, for instance, to the nine stars in the belt and sword of Orion he was able to add eighty others which he discovered in their vicinity; and, to the seven in the Pleiades, another thirty-six. The Milky Way dissolved before the telescope into "a mass of innumerable stars

duced no important argument in favor of Copernicus, nor any clear committal on his part. Besides, the discoveries announced in the *Star Messenger* were not quite as original as they pretended to be. He was neither the first nor the only scientist who had turned a telescope at the sky and discovered new wonders with it. Thomas Harriot made systematic telescopic observations and maps of the moon in the summer of 1609, before Galileo, but he did not publish them. Even the Emperor Rudolph had watched the moon through a telescope before he had heard of Galileo. Galileo's star maps were so inaccurate that the Fleiades group can only be identified on them with difficulty, the Orion group not at all; and the huge dark spot under the moon's equator, surrounded by mountains, which Galileo compared to Bohemia, simply does not exist.

Yet when all this is said, and all the holes are picked in Galileo's first published text, its impact and significance still remain tremendous. Others had seen what Galileo saw, and even his priority in the discovery of the Jupiter moons is not established beyond doubt[18a]; yet he was the first to publish what he saw, and to describe it in a language which made everybody sit up. It was the cumulative effect that made the impact; the vast philosophical implications of this further prising open of the universe were instinctively felt by the reader, even if they were not explicitly stated. The mountains and valleys of the moon confirmed the similarity between heavenly and earthly matter, the homogeneous nature of the stuff from which the universe is built. The unsuspected number of invisible stars made an absurdity of the notion that they were created for man's pleasure, since he could see them only armed with a machine. The Jupiter moons did not prove that Copernicus was right, but they did further shake the antique belief that the earth was the center of the world around which everything turned. It was not this or that particular detail but the total contents of the *Messenger from the Stars* that created the dramatic effect.

The booklet aroused immediate and passionate controversy. It is curious to note that Copernicus' *Book of Revolutions* had created little stir for half a century, and Kepler's Laws even less at their time, while the *Star Messenger*, which had only an indirect bearing on the issue, caused such an outburst of emotions. The main reason was, no doubt, its immense readability. To digest Kepler's magnum opus required, as one of his colleagues remarked, "nearly a lifetime"; but the *Star Messenger* could be read in an hour, and its effect was like a punch in the solar plexus on those grown up in the traditional view of the bounded universe. And that vision, though a bit shaky, still retained an immense, reassuring coherence.

Even Kepler was frightened by the wild perspective opened up by Galileo's spyglass. "The infinite is unthinkable!" he repeatedly exclaimed in anguish.

The shock waves of Galileo's message spread immediately, as far as England. It was published in March 1610; Donne's *Ignatius* was published barely ten months later,[19] but Galileo (and Kepler) are repeatedly mentioned in it.

I will write [quoth Lucifer] to the Bishop of Rome:
He shall call Galileo the Florentine to him . . .

But soon, the satirical approach yielded to the metaphysical, to a full realization of the new cosmic perspective.

Man has weav'd out a net, and this net throwne
Upon the Heavens, and now they are his owne . . .

Milton was still an infant in 1610; he grew up with the new wonders. His awareness of the "vast unbounded Deep" that the telescope disclosed reflects the end of the medieval walled universe.

Before [his] eyes in sudden view appear
The secrets of the hoary Deep—a dark
Illimitable ocean, without bound,
Without dimension . . .[20]

6. THE BATTLE OF THE SATELLITES

Such was the objective impact on the world at large of
Galileo's discoveries with his "optick tube." But to understand the reactions of the small academic world in
his own country, we must also take into account the subjective effect of Galileo's personality. Copernicus had
been a kind of invisible man throughout his life; nobody
who met the disarming Kepler, in the flesh or by correspondence, could seriously dislike him. But Galileo had
a rare gift of provoking enmity—not the affection alter-
nating with rage which Tycho aroused, but the cold, un-
relenting hostility which genius plus arrogance minus
humility creates among mediocrities.

Without this personal background, the controversy
that followed the publication of the *Sidereus Nuncius*
would remain incomprehensible. For the subject of the
quarrel was not the *significance* of the Jupiter satellites,
but their *existence*, which some of Italy's most illustrious scholars flatly denied. Galileo's main academic
rival was Magini in Bologna. In the month following the
publication of the *Star Messenger*, on the evenings of
April 24 and 25, 1610, a memorable party was held in
a house in Bologna, where Galileo was invited to demonstrate the Jupiter moons in his spyglass. Not one among
the numerous and illustrious guests declared himself
convinced of their existence. Father Clavius, the leading
mathematician in Rome, equally failed to see them;
Cremonini, teacher of philosophy at Padua, refused even
to look into the telescope; so did his colleague Libri.
The latter, incidentally, died soon afterward, providing
Galileo with an opportunity to make more enemies with

the much-quoted sarcasm: "Libri did not choose to see my celestial trifles while he was on earth; perhaps he will do so now he has gone to heaven."

These men may have been partially blinded by passion and prejudice, but they were not quite as stupid as it may seem. Galileo's telescope was the best available, but it was still a clumsy instrument without fixed mountings, and with a visual field so small that, as somebody has said, "the marvel is not so much that he found Jupiter's moons, but that he was able to find Jupiter itself." The tube needed skill and experience in handling, which none of the others possessed. Sometimes a fixed star appeared in duplicate. Moreover, Galileo himself was unable to explain why and how the thing worked; and the *Sidereus Nuncius* was conspicuously silent on this essential point. Thus it was not entirely unreasonable to suspect that the blurred dots which appeared to the strained and watering eye pressed to the spectacles-sized lens might be optical illusions in the atmosphere, or somehow produced by the mysterious gadget itself. This, in fact, was asserted in a sensational pamphlet, *Refutation of the Star Messenger*,[20a] published by Magini's assistant, a young fool called Martin Horky.

Thus, while the poets were celebrating Galileo's discoveries, which had become the talk of the world, the scholars in his own country were, with very few exceptions, hostile or skeptical. The first, and for some time the only, scholarly voice raised in public in defense of Galileo was Johannes Kepler's.

7. THE SHIELD BEARER

It was also the weightiest voice, for Kepler's authority as the first astronomer of Europe was uncontested—not because of his two Laws, but by virtue of his position as imperial mathematicus and successor to Tycho. John Donne, who had a grudging admiration for him, has summed up Kepler's reputation "who (as him-selfe testifies of himselfe) ever since Tycho Brahe's death hath received it into his care, that no new thing should be done in heaven without his knowledge."[21]

The first news of Galileo's discovery had reached Kep-ler when Wackher von Wackenfeld called on him on or around March 15, 1610. The weeks that followed he spent in feverish expectation of more definite news. In the first days of April, the Emperor received a copy of the *Star Messenger*, which had just been published in Venice, and Kepler was graciously permitted "to have a look and rapidly glance through it." On April 8, at last, he received a copy of his own from Galileo, accompanied by a request for his opinion.

Galileo had never answered Kepler's fervent request for an opinion on the *Mysterium*, and had remained equally silent on the *New Astronomy*. Nor did he bother to put his own request for Kepler's opinion on the *Star Messenger* into a personal letter. It was transmitted to Kepler verbally by the Tuscan ambassador in Prague, Julian de Medici. Although Kepler was not in a position to verify Galileo's disputed discoveries, for he had no telescope, he took Galileo's claims on trust. He did it enthusiastically and without hesitation, publicly offering to serve in the battle as Galileo's "squire" or "shield bearer"—he, the imperial mathematicus, to the recently still unknown Italian scholar. It was one of the most generous gestures in the annals of science.

The courier for Italy was to leave on April 19; in the eleven days at his disposal Kepler wrote his pamphlet, *Conversation with the Star Messenger*, in the form of an open letter to Galileo. It was printed the next month in Prague, and a pirated Italian translation appeared shortly afterward in Florence.

It was precisely the support that Galileo needed at that moment. The weight of Kepler's authority played an important part in turning the tide of the battle in his favor, as shown by Galileo's correspondence. He was anxious to leave Padua and to be appointed court mathematician to Cosimo de Medici, Grand Duke of Tuscany, in whose honor he had called Jupiter's planets "the Medicean stars." In his application to Vinta, the duke's secretary of state, Kepler's support figures prominently.

"Your Excellency, and their Highnesses through you, should know that I have received a letter—or rather an eight-page treatise—from the Imperial Mathematician, written in approbation of every detail contained in my book without the slightest doubt or contradiction of anything. And you may believe that this is the way leading men of letters in Italy would have spoken from the beginning if I had been in Germany or somewhere far away."[22]

He wrote in almost identical terms to other correspondents, among them to Matteo Carosio in Paris.

"We were prepared for it that twenty-five people would wish to refute me; but up to this moment I have seen only one statement by Kepler, the Imperial Mathematician, which confirms everything that I have written, without rejecting even an iota of it; which statement is now being reprinted in Venice, and you shall soon see it."[23]

Yet, while Galileo boasted about Kepler's letter to the grand duke and his correspondents, he neither thanked Kepler nor even acknowledged it.

Apart from its strategical importance in the cosmological battle, the *Conversation with the Star Messenger* is without much scientific value; it reads like a baroque arabesque, a pattern of amusing doodles around the hard core of Galileo's treatise. It starts with Kepler's voicing his hope that Galileo, whose opinion matters to him more than anybody's, would comment on the *Astronomia Nova*, and thereby renew a correspondence "laid aside twelve years ago." He relates with gusto how he had received the first news of the discoveries from Wackher—and how he had worried whether the Jupiter moons could be fitted into the universe built around the five Pythagorean solids. But as soon as he had cast a glance at the *Star Messenger*, he realized that "it offered a highly important and wonderful show to astronomers and philosophers, that it invited all friends of true philosophy to contemplate matters of the highest import. . . . Who could be silent in the face of such a message? Who would not feel himself overflow with the love of the Divine which is so abundantly manifested here?" Then comes his offer of support "in the battle against the grumpy reactionaries, who reject everything that is unknown as unbelievable, and regard everything that departs from the beaten track of Aristotle as a desecration. . . . Perhaps I shall be considered reckless because I accept your claims as true without being able to add my own observations. But how could I distrust a reliable mathematician whose art of language alone demonstrates the straightness of his judgment? . . ."

Kepler had instinctively felt the ring of truth in the *Star Messenger*, and that had settled the question for him. However much he may have resented Galileo's previous behavior, he felt committed "to throw himself into the fray" for Truth, Copernicus, and the Five Perfect Solids. For, having finished the Promethean labors of the *New Astronomy*, he was again steeped in the mystic twilight of a Pythagorean universe built around cube, tetrahedra, dodecahedra, and so on. They are the leitmotif of his dialogue with the *Star Messenger*; not even once did he mention the elliptical orbits, the First or the Second Law. Their discovery appeared to him merely as a tedious detour in the pursuit of his *idée fixe*.

It is a rambling treatise, written by a hurried pen that jumps from one subject to another: astrology, optics, the moon's spots, the nature of the ether, Copernicus, the habitability of other worlds, interplanetary travel.

"There will certainly be no lack of human pioneers when we have mastered the art of flight. Who would have thought that navigation across the vast ocean is less dangerous and quieter than in the narrow, threatening gulfs of the Adriatic, or the Baltic, or the British straits? Let us create vessels and sails adjusted to the heavenly ether, and there will be plenty of people unafraid of the empty wastes. In the meantime, we shall prepare, for the brave sky travelers, maps of the celestial bodies—I shall do it for the moon, you, Galileo, for Jupiter."

Living in an atmosphere saturated with malice, Professors Magini, Horky, and even Maestlin could not believe their ears when they heard Kepler singing Galileo's praises, and they tried to discover some hidden sting in the treatise. They gloated over a passage in which Kepler showed that the principle of the telescope had been outlined twenty years before by one of Galileo's countrymen, Giovanni Della Porta, and by Kepler himself in his work on optics in 1604. But since Galileo did not claim the invention of the telescope, Kepler's historical excursion could not be resented by him; moreover, Kepler emphasized that Della Porta's and his own anticipations were of a purely theoretical nature "and cannot diminish the fame of the inventor, whoever it was. For I know what a long road it is from a theoretical concept to its practical achievement, from the mention of the antipodes in Ptolemy to Columbus' discovery of the New World, and even more from the two-lensed instruments used in this country to the instrument with which you, O Galilee, penetrated the very skies."

In spite of this, the German envoy in Venice, Georg Fugger, wrote with relish that Kepler had "torn the mask off Galileo's face,"[24] and Francis Stelluti (a member of the Lincean Academy) wrote to his brother, "According to Kepler, Galileo makes himself out to be the inventor of the instrument, but more than thirty years ago Della Porta described it in his *Natural Magic.* . . . And so poor Galileo will look foolish."[25] Horky also quoted Kepler in his much-read pamphlet against Galileo, whereupon Kepler immediately informed Horky that "since the demands of honesty have become incompatible with my friendship for you, I hereby terminate the latter,"[26] and offered Galileo to publish the rebuke; but when the youngster relented, he forgave him.

These reactions indicate the extent of dislike for Galileo in his native Italy. But whatever hidden irony the scholars had imputed to Kepler's *Dissertatio*, the undeniable fact was that the imperial mathematicus had expressly endorsed Galileo's claims. This persuaded some of Galileo's opponents, who had previously refused to take him seriously, to look for themselves through improved telescopes which were now becoming available. The first among the converts was the leading astronomer in Rome, the Jesuit Father Clavius. In the sequel, the Jesuit scholars in Rome not only confirmed Galileo's observations, but considerably improved on them.

8. THE PARTING OF THE ORBITS

Galileo's reaction to the service Kepler had rendered him was, as we saw, complete silence. The Tuscan ambassador at the imperial court urgently advised him to send Kepler a telescope to enable him to verify, at least *post factum*, Galileo's discoveries, which he had accepted on trust. Galileo did nothing of the sort. The telescopes his workshop turned out he donated to various aristocratic patrons.

Four months thus went by, Horky's pamphlet was published, the controversy had reached its peak, and so far not a single astronomer of repute had publicly confirmed having seen the moons of Jupiter. Kepler's friends began to reproach him for having testified to what he himself had not seen; it was an impossible situation.[26a] On August 9, he again wrote to Galileo.

". . . You have aroused in me a great desire to see your instrument so that at last I too can enjoy, like yourself, the spectacle of the skies. For among the instruments at our disposal here the best magnifies only ten times, the others hardly thrice. . . ."[27]

He talked about his own observations of Mars and the moon, expressed his indignation at Horky's knavery; and then continued:

"The law demands that everybody should be trusted unless the contrary is proven. And how much more is this the case when the circumstances warrant trustworthiness. In fact, we are dealing not with a philosophical but with a legal problem: did Galileo deliberately mislead the world by a hoax? . . .

I do not wish to hide from you that letters have reached Prague from several Italians who deny that those planets can be seen through your telescope.

I am asking myself how it is possible that so many deny [their existence], including those who possess a telescope. . . . Therefore I ask you, my Galileo, nominate witnesses for me as soon as possible. From various letters written by you to third persons I have learned that you do not lack such witnesses. But I am unable to name any testimony except your own. . . ."[27a]

This time Galileo hurried to answer, evidently scared by the prospect of losing his most powerful ally.

"Padua, August 19, 1610.
I have received both your letters, my most

learned Kepler. The first, which you have already published, I shall answer in the second edition of my observations. In the meantime, I wish to thank you for being the first, and almost the only, person who completely accepted my assertions, though you had no proof, thanks to your frank and noble mind."[28]

Galileo went on to tell Kepler that he could not lend him his telescope, which magnified a thousandfold, because he had given it to the grand duke, who wished "to exhibit it in his gallery as an eternal souvenir among his most precious treasures." He made various excuses about the difficulty of constructing instruments of equal excellence, ending with the vague promise that he would, as soon as possible, make new ones "and send them to my friends." Kepler never received one.

In the next paragraph Horky and the vulgar crowd came in for some more abuse—"but Jupiter defies both giants and pygmies; Jupiter stands in the sky, and the sycophants may bark as they wish." Then he turned to Kepler's request for witnesses, but still could not name a single astronomer. "In Pisa, Florence, Bologna, Venice, and Padua, a good many have seen [the Medicean stars] but they are all silent and hesitate." Instead, he named his new patron, the grand duke, and another member of the Medici family (who could hardly be expected to deny the existence of stars named after them). He continued:

"As a further witness I offer myself, who have been singled out by our university for a lifelong salary of a thousand florins, such as no mathematician has ever enjoyed, and which I would continue to receive forever even if the Jupiter moons were to deceive us and vanish."

After complaining bitterly about his colleagues "most of whom are incapable of identifying either Jupiter or

Mars, and hardly even the moon," Galileo concluded:

"What is to be done? Let us laugh at the stupidity of the crowd, my Kepler. . . . I wish I had more time to laugh with you. How you would shout with laughter, my dearest Kepler, if you were to hear what the chief philosophers of Pisa said against me to the grand duke. . . . But the night has come and I can no longer converse with you. . . ."

This is the second, and last, letter Galileo ever wrote to Kepler.[29] The first, it will be remembered, was written thirteen years earlier, and its theme song had been the perversity of philosophers and the stupidity of the crowd, concluding with the wistful remark, "if only more people like Kepler existed." Now, writing for the first time after these thirteen years, he again singled out Kepler as a unique ally to laugh with him at the foolishness of the world. But concerning the quandary into which his loyal ally had got himself, the letter was as unhelpful as could be. It contained not a word on the progress of Galileo's observations, about which Kepler was burning to hear; and it made no mention of an important new discovery that Galileo had made, and that he had communicated, about a fortnight earlier, to the Tuscan ambassador in Prague.[30] The communication ran as follows:

"SMAISMRMILMEPOETALEUMIBUNENUGTTAURIAS."

This meaningless sequence of letters was an anagram made up from the words describing the new discovery. The purpose behind it was to safeguard the priority of the find without disclosing its content, lest somebody else might claim it as his own. Ever since the affair of the proportional compass, Galileo had been very anxious to secure the priority of his observations—even, as we shall see, in cases where the priority was not his. But whatever his motives in general, they can hardly excuse the fact that he asked the Tuscan ambassador to dangle the puzzle before the tantalized eyes of Kepler, whom he could not suspect of intending to steal his discovery.

Poor Kepler tried to solve the anagram, and patiently transformed it into what he himself called a "barbaric Latin verse": "*Salve umbistineum geminatum Martia proles*"—"Hail, burning twin, offspring of Mars."[31] He accordingly believed that Galileo had discovered moons around Mars too. Only three months later, on November 13, did Galileo condescend to disclose the solution —not, of course, to Kepler, but to Rudolph, because Julian de Medici informed him that the Emperor's curiosity was aroused.

The solution was: "*Altissimum planetam tergeminum observavi*"—"I have observed the highest planet [Saturn] in triplet form." Galileo's telescope was not powerful enough to disclose Saturn's rings (they were only seen half a century later by Huygens); he believed Saturn to have two small moons on opposite sides, and very close to the planet.

A month later, he sent another anagram to Julian de Medici: "*Haec immatura a me jam frustra legunturoy*" —"These immature things I am searching for now in vain." Once again Kepler tried several solutions, among them: "*Macula rufa in Jove est gyratur mathem, etc.*";* then wrote to Galileo in exasperation:

"I beseech you not to withhold from us the solution for long. You must see that you are dealing with honest Germans . . . consider what embarrassment your silence causes me."[32]

Galileo disclosed his secret a month later—again not directly to Kepler, but to Julian de Medici: "*Cynthiae figuras aemulatur mater amorum*"—"The mother of love [Venus] emulates the shapes of Cynthia [the moon]." Galileo had discovered that Venus, like the moon,

* "There is a red spot in Jupiter which rotates mathematically."

Ioannis Keppleri

HARMONICES
MVNDI

LIBRI V. Qvorvm

Primus Geometricvs, De Figurarum Regularium, quæ Proportiones Harmonicas conftituunt, ortu & demonftrationibus.

Secundus Architectonicvs, feu ex Geometria Figvrata, De Figurarum Regularium Congruentia in plano vel folido:

Tertius proprie Harmonicvs, De Proportionum Harmonicarum ortu ex Figuris; deque Natura & Differentiis rerum ad cantum pertinentium, contra Veteres:

Quartus Metaphysicvs, Psychologicvs & Astrologicvs, De Harmoniarum mentali Effentia earumque generibus in Mundo; præfertim de Harmonia radiorum, ex corporibus cœleftibus in Terram defcendentibus, eiufque effectu in Natura feu Anima fublunari & Humana:

Quintus Astronomicvs & Metaphysicvs, De Harmoniis abfolutiffimis motuum cœleftium, ortuque Eccentricitatum ex proportionibus Harmonicis.

Appendix habet comparationem huius Operis cum Harmonices Cl. Ptolemæi libro III.cumque Roberti de Fluctibus, dicti Flud.Medici Oxonienfis fpeculationibus Harmonicis, operi de Macrocofmo & Microcofmo infertis.

Cum S.C.Mᵗⁱ. Priuilegio ad annos XV.

Lincii Auftriæ,

Sumptibus Godofredi Tampachii Bibl. Francof.
Excudebat Ioannes Plancvs.

Anno M. DC. XIX.

PLATE V. Title page of *Harmonices Mundi* (1619).

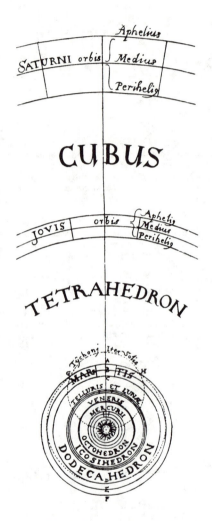

PLATE VI. Kepler's diagram of the planets' orbits inside the five regular solids of Plate I. From *Harmonices Mundi*, Liber V (1619).

Prodromus

DISSERTATIONVM COSMOGRAPHICARVM,

continens

MYSTERIVM
COSMOGRAPHICVM

DE ADMIRABILI PROPORTIONE OR-
bium cœleſtium: deque cauſis cœlorum numeri, magni-
tudinis, motuumque periodicorum ge-
nuinis & propriis,

Demonſtratum per quinque regularia corpora Geometrica.

Libellus primum Tübingæ in lucem datus Anno Chriſti
M. D X C V I.

à

*M. IOANNE KEPLERO VVIRTEMBERGICO, TVNC TEMPO-
ris Illuſtrium Styriæ Prouincialium Mathematico.*

Nunc vero poſt annos 25. ab eodem authore recognitus, & Notis notabiliſſimis
partim emendatus, partim explicatus, partim confirmatus : deniq; omnibus ſuis
membris collatus ad alia cognati argumenti opera, quæ Author ex illo tem-
pore ſub duorum Impp. Rudolphi & Matthiæ auſpiciis; etiamq; in
Illuſtr. Ord. Auſtriæ Supr-Aniſanæ clientela
diuerſis locis edidit.

*Potiſſimum ad illuſtrandas occaſiones Operis, Harmonice Mundi, dicti, eiuſ-
que progreſſuum in materia & methodo.*

Addita eſt erudita NARRATIO M. GEORGII IOACHIMI RHETICI, de
Libris Reuolutionum, atque admirandis de numero, ordine, & diſtantiis Sphæra-
rum Mundi hypotheſibus, excellentiſſimi Mathematici, totiuſque Aſtronomiæ Re-
ſtauratoris D. NICOLAI COPERNICI.

ITEM,

Eiuſdem IOANNIS KEPLERI pro ſuo Opere Harmonices Mundi APOLOGIA aduer-
ſus Demonſtrationem Analyticam Cl. V. D. Roberti de Fluctibus , Medici Oxonienſis

Cum Priuilegio Cæſareo ad annos XV

FRANCOFVRTI,
Recuſus Typis ERASMI KEMPFERI , ſumptibus
GODEFRIDI TAMPACHII.

Anno M. DC. XXI.

PLATE VII. Title page of second edition of
Mysterium Cosmographicum (1621).

Ioannis Kepleri

MATHEMATICI,
PRO SVO OPERE HARMO-
NICES MVNDI
APOLOGIA.

ADVERSVS DEMONSTRATIO-
nem Analyticam CL. V. D. Roberti de Fluctibus
Medici Oxonienſis.

*IN QVA ILLE SE DICIT RESPONDERE
ad Appendicem dicti Operis.*

FRANCOFVRTI
Sumptibus GODEFRIDI TAMPACHII.

Anno M. DC. XXII.

PLATE VIII. Title page of Apologia (1622).
Found in *Mysterium Cosmographicum*
(later printing of 1621 edition).

showed phases—from sickle to full disk and back—a proof that she revolved around the sun. He also considered this as proof of the Copernican system—which it was not, for it equally fitted the Egyptian or the Tychonic system. In the meantime, Kepler's dearest wish, to see for himself the new marvels, was at last fulfilled. One of Kepler's patrons, the Elector Ernest of Cologne, Duke of Bavaria, was among the select few whom Galileo had honored with the gift of a telescope. In the summer of 1610 Ernest was in Prague on affairs of state, and for a short period lent his telescope to the imperial mathematicus. Thus from August 3 to September 9 Kepler was able to watch the Jupiter moons with his own eyes. The result was another short pamphlet, *Observation Report on Jupiter's Four Wandering Satellites*,[33] in which Kepler confirmed, this time from firsthand experience, Galileo's discoveries. The treatise was immediately reprinted in Florence, and was the first public testimony by independent, direct observation, of the existence of the Jupiter moons. It was also the first appearance in history of the term "satellite," which Kepler had coined in a previous letter to Galileo.[34]

At this point the personal contact between Galileo and Kepler ends. For a second time Galileo broke off their correspondence. In the subsequent months, Kepler wrote several more letters, which Galileo left unanswered, or answered indirectly by messages via the Tuscan ambassador. Galileo wrote to Kepler only once during this whole period of the "meeting of their orbits": the letter of August 19, 1610, which I have quoted. In his works he rarely mentions Kepler's name, and mostly with intent to refute him. Kepler's three Laws, his discoveries in optics, and the Keplerian telescope are ignored by Galileo, who firmly defended to the end of his life circles and epicycles as the only conceivable form of heavenly motion.

chapter nine

Chaos and Harmony

1. "DIOPTRICE"

Galileo had transformed the Dutch spyglass from a toy into an instrument of science, but he had nothing to say in explanation of why and how it worked. It was Kepler who did this. In August and September 1610, while he enjoyed the use of the telescope borrowed from Duke Ernest of Cologne, he wrote within a few weeks a theoretical treatise in which he founded a new science and coined a name for it: dioptrics—the science of refraction by lenses. His *Dioptrice*[1] is a classic of a strikingly un-Keplerian kind, consisting of a hundred and forty-one austere "definitions," "axioms," "problems," and "propositions" without any arabesque, ornament, or mystic flights of thought.[2] Though he did not find the precise formulation of the law of refraction, he was able to develop his system of geometrical and instrumental optics, and to deduce from it the principles of the so-called astronomical, or Keplerian telescope.

In his previous book on optics, published in 1604, Kepler had shown that the intensity of light diminishes with the square of distance; he had explained the principle of the *camera obscura*, the forerunner of the photographic camera, and the manner in which the spectacles for the short- and longsighted worked. Spectacles had been in use since antiquity, but there existed no precise theory for them. Nor, if it comes to that, did a

satisfactory explanation exist for the process of sight—
the refraction of the incoming light by the lenses in the
eye, and the projection of a reversed image onto the
retina—until Kepler's first book on optics. He had modestly called it "a Supplement to Vitellio."[3] This Vitellio,
a thirteenth-century scholar, had written a compendium
of optics mainly based on Ptolemy and Alhazen, and this
was the most up-to-date work on the subject till Kepler's
advent. One must constantly bear in mind this lack of
continuity in the development of science, the immense,
dark lowlands extending between the peaks of antiquity
and the watershed, to see the achievements of Kepler
and Galileo in true perspective.

The *Dioptrice* is Kepler's soberest work—as sober as
the geometry of Euclid. He wrote it in the same year
as his punch-drunk *Conversation with the Star Messenger*. It had been one of the most exciting years in Kepler's life; it was followed by the blackest and most
depressing.

2. DISASTER

The year 1611 brought civil war and epidemics to
Prague; the abdication of his imperial patron and pro-
vider; the death of his wife and favorite child.

Men less prone to astrology would have blamed such
a series of catastrophes on the evil influence of the stars;
oddly enough, Kepler did not. His astrological beliefs
had become too refined for that: he still believed that
the constellations influenced the formation of character,
and also had a kind of catalyzing effect on events; but
the cruder form of direct astrological causation he rejected as superstition.

This made his position at court even more difficult.
Rudolph, sliding from apathy into insanity, was now vir-
tually a prisoner in his citadel. His cousin Leopold had
raised an army and occupied part of Prague. The Bohe-

mian Estates appealed for help to his brother Matthias, who had already dispossessed Rudolph of Austria, Hungary, and Moravia, and was preparing to take over what was left. Rudolph craved reassurance from the stars; but Kepler was too honest to provide it. In a confidential letter to one of Rudolph's intimate advisers, he explained:

> "Astrology can do enormous harm to a monarch if a clever astrologer exploits his human credulity. I must watch that this should not happen to our Emperor. . . . I hold that astrology must not only be banished from the Senate, but also from the heads of all who wish to advise the Emperor in his best interests; it must be kept entirely out of his sight."[4]

He went on to say that, consulted by the Emperor's enemies, he had pretended that the stars were favorable to Rudolph and unfavorable to Matthias; but he would never say this to the Emperor himself, lest he became overconfident and neglected whatever chance there may be left to save his throne. Kepler was not above writing astrological calendars for money, but where his conscience was involved, he acted with a scrupulousness most unusual by the standards of his time.

On May 23, Rudolph was forced to abdicate the Bohemian crown; the following January he was dead. In the meantime, Frau Barbara contracted the Hungarian fever, which was followed by attacks of epilepsy and symptoms of mental derangement. When she got better, the three children went down with the pox, which the soldiery had imported. The oldest and youngest recovered; the favorite, six-year-old Friedrich, died. Then Barbara relapsed.

> "Numbed by the horrors committed by the soldiers, and the bloody fighting in the town; con-

sumed by despair of the future and by an un-
quenchable longing for her lost darling . . . in
melancholy despondency, the saddest of all states
of mind, she gave up the ghost."[5]

It was the first in a series of disasters which weighed
down on the last twenty years of Kepler's life. To keep
going, he published his correspondence with various
scholars on questions of chronology in the age of Christ.
Chronology had always been one of his favorite distrac-
tions; his theory that Jesus was really born in the year
4 or 5 "B.C." is today generally accepted. Thus he was
"marking time" in two meanings of the word; for he had
secured himself a new, modest job in Linz, but could
not leave Prague while Rudolph was still alive.

The end came on January 20, 1612. It was also the
end of the most fertile and glorious period in Kepler's
life.

3. EXCOMMUNICATION

The new job was that of a provincial mathematicus
in Linz, capital of Upper Austria—similar to that he had
held in his youth in Gratz. He was now forty-one years
old, and he stayed in Linz for fourteen years, till he was
fifty-five.

It seemed a depressing comedown after the glories of
Prague; but it was not quite as bad as it seemed. For
one thing, Rudolph's successor had confirmed Kepler in
his title as imperial mathematician, which he retained
throughout his life. Matthias, unlike Rudolph, had little
time for his court astronomer; but he wanted him to
be not too far away, and Linz, in his Austrian domain,
was a satisfactory solution. Kepler himself was glad to
be away from the turmoil of Prague, and to receive a
salary from the Austrians which at least he was sure to
get. He also had influential patrons among the local

aristocracy, the Starhembergs and Liechtensteins—in fact, the job had been specially created for him, carried only theoretical obligations, and left him all the leisure he needed for his work. When the Thirty Years' War began with the defenestration of Prague, he could only be thankful to be removed from the focus of events. And when he was offered the succession of Magini in the Chair of Mathematics at Bologna, he wisely refused.

But nevertheless it was a comedown. "Linz," to Austrians, remains to this day a byword for provincialism. Barbara, whose homesickness for Austria had been one of the reasons for Kepler's choice of Linz, was dead. His desolate loneliness wrung from him one of his self-analytical outcries:

> ". . . My exaggerated trustingness, display of piety, a clutching at fame by means of startling projects and unusual actions, the restless search for and interpretation of causes, the spiritual anguish for grace. . . ."[6]

He had nobody to talk to, nobody even to quarrel with.

This last need, however, was fulfilled after a while by the local parson, one Daniel Hitzler. He also came from Wuerttemberg, and knew all about Kepler's scandalous crypto-Calvinist deviations. On the first occasion when Kepler came for Communion, they had an argument. Kepler denied, as he had always done, the Lutheran doctrine of the ubiquity—the omnipresence in the world—not only of the spirit, but of the body of Christ; while Hitzler insisted on a written statement of conformity to the doctrine (which, later on, was dropped by Lutheran theology). Kepler refused, whereupon Hitzler refused him Communion. Kepler complained in a fervent petition to the Church council in Wuerttemberg; the council answered in a long, patient, and paternally chiding letter that Kepler should stick to mathematics and leave theology to the theologians. Kepler was forced to go for Communion to a parish outside Linz, whose parson was apparently more broad-minded; the Church council, while backing Pastor Hitzler, did nothing to prevent his colleague from giving Communion to the errant sheep. Kepler kept protesting against the curtailment of his freedom of conscience, and complaining that gossips called him an atheist and a double-dealer who was trying to curry favor with the Catholics and flirting with the Calvinists. Yet this repeated falling between three stools seemed to agree with his innermost nature.

"It hurts my heart that the three factions have miserably torn the truth to pieces between them, that I must collect the bits wherever I can find them, and put them together again. . . . I labor to reconcile the parties with each other whenever it can be done with sincerity, so that I should be able to side with all of them. . . . Behold, I am attracted either by all three parties, or at least by two of them against the third, setting my hopes on agreement; but my opponents are only attracted by one party, imagining that there must be irreconcilable division and strife. My attitude, so help me God, is a Christian one; theirs, I do not know what."[7]

It was the language of Erasmus and Tiedemann Giese, of the golden age of tolerance—but entirely out of place and out of date in Germany on the eve of the Thirty Years' War.

Engulfed in that European disaster, Kepler had to endure an additional ordeal: a kind of ghastly private epicycle turning on the greater wheel. His old mother had been accused of witchcraft and was threatened with being burned alive. The proceedings lasted for six years, from 1615 to 1621; compared with this, the quasi- or semi-excommunication of Kepler himself was only a minor nuisance.

4. THE WITCH TRIAL

The witch-hunting mania, which had grown in furor throughout the sixteenth century, reached its peak in the first half of the seventeenth, both in the Catholic and Protestant parts of Germany. In Weil-der-Stadt, Kepler's idyllic birthplace, with a population of two hundred families, thirty-eight witches were burned between 1615 and 1629. In the neighboring Leonberg, where Kepler's mother now lived, a place equally small, six witches were burned in the winter of 1615 alone. It was one of the hurricanes of madness which strike the world from time to time, and seem to be part of man's condition.

Kepler's mother was now a hideous little old woman, whose meddlesomeness and evil tongue, together with her suspect background, predestined her as a victim. She was, we remember, an innkeeper's daughter, brought up by an aunt who was said to have perished at the stake; and her husband had been a mercenary who vanished after barely escaping the gallows. In that same year, 1615, when Leonberg was seized with witch hysteria, Katherine had a quarrel with another old hag, her former best friend, the wife of the glazier Jacob Reinhold. This was to be her undoing. The glazier's wife accused Katherine of having given her a witches' potion which had produced a chronic illness (in fact, her ailment was caused by an abortion). It was now remembered that various burghers of Leonberg had been taken ill at various times after being offered a drink from a tin jug which Katherine always kept hospitably prepared for her visitors. The wife of Bastian Meyer had died of it, and the schoolmaster Beutelspacher was permanently paralyzed. It was remembered that once upon a time Katherine had asked the sexton for the skull of her father, which she wanted to have cast in silver as a drinking goblet for her son—that court astrologer, himself an adept of the black art. She had cast an evil eye on the children of the tailor Daniel Schmidt, who had promptly died; she was known to have entered houses through locked doors, and to have ridden to death a calf, of which she offered a cutlet to her other son, Heinrich the vagrant.

Katherine's foremost enemy, the glazier's wife, had a brother, who was court barber to the Duke of Wuerttemberg. In that fateful year, 1615, the duke's son, Prince Achilles, came to Leonberg to hunt, with the barber in his suite. The barber and the town provost got drunk together, and had Ma Kepler brought to the town hall. Here the barber put the point of his sword at the old woman's breast and asked her to cure his sister by witches' magic of the ailment she had cast on her. Katherine had the sense to refuse—otherwise she would have convicted herself—and her family now sued for libel to protect her. But the town provost blocked the libel suit by starting formal proceedings against Katherine for witchcraft. The incident which provided him with an opportunity to do so involved a girl of twelve, who was carrying bricks to the kiln, and on passing Ma Kepler on the road felt a sudden pain in the arm, which led to a temporary paralysis. These sudden, stabbing pains in shoulder, arm, or hip played a great part in the trial of Katherine and others; to this day, lumbago pains and stiff necks are called in Germany "*Hexenschuss*"— witches' shot.

The proceedings were long, ghastly, and squalid. At various stages, Kepler's younger brother, Christoph, drillmaster of the militia of Leonberg, and his brother-in-law, the vicar, dissociated themselves from the old woman, squabbled over the cost of defense, and would apparently have been quite glad to see their mother burned and have done with, except for the reflection it would cast on their own bourgeois respectability. Kepler had always been fated to fight without allies, and for unpopular causes. He started with a counterattack, accusing his mother's persecutors of being inspired by the devil, and peremptorily advised the town council of Leonberg to watch their steps, to remember that he was his Roman Imperial Majesty's Court Mathematicus, and to send him copies of all documents relating to his mother's case. This opening blast had the desired effect of making the town provost, the barber, and their clique proceed more warily, and to look for more evidence before applying for formal indictment. Ma Kepler obligingly provided it, by offering the provost a silver goblet as a bribe if he consented to suppress the report on the incident of the little girl with the bricks. After that, her son, daughter, and son-in-law decided that the only solution was flight, and bundled Ma Kepler off to Johannes in Linz, where she arrived in December 1616. This done, Christoph and the vicar wrote to the ducal chancellery that should the provost's accusations prove justified, they would disown old Katherine and let justice take its course.

The old woman stayed for nine months in Linz; then she got homesick and returned to live with Margaret and the vicar, stake or no stake. Kepler followed her, reading on the journey *The Dialogue on Ancient and Modern Music* by Galileo's father. He stayed in Wuerttemberg for two months, wrote petitions, and tried to obtain a hearing of the original libel suit—to no avail. He succeeded only in obtaining permission to take his mother back with him to Linz. But the stubborn old woman refused; she did not like Austria. Kepler had to return without her.

There followed a strange lull of two years—the opening years of the Thirty Years' War—during which Kepler wrote more petitions and the court collected more evidence, which now filled several volumes. Finally, on the night of August 7, 1620, Ma Kepler was arrested in her son-in-law's vicarage; to avoid scandal, she was carried out of the house, hidden in an oak linen chest, and thus transported to the prison in Leonberg. She was interrogated by the provost, denied being a witch, and was committed to a second and last interrogation, before being put to the torture.

Margaret sent another SOS to Linz, and Kepler set out once again for Wuerttemberg. The immediate result of his arrival was that the supreme court granted Ma Kepler six weeks to prepare her defense. She was lying in chains in a room at the town gate, with two full-time guards—whose salary had to be paid by the defense, in addition to the cost of the extravagant quantities of firewood they burned. Kepler, who had built a new astronomy on a trifle of eight minutes' arc, did not neglect such details in his petitions; he pointed out that one guard would be a sufficient security precaution for his chained mother, aged seventy-three, and that the expense of the firewood should be more equitably shared. He was his irrepressible, indefatigable, passionate, and precise self. The situation, from the point of view of the authorities, was summed up by a slip in the court scribe's record: "The accused appeared in court, accompanied, alas, by her son, Johannes Kepler, mathematician."[8]

The proceedings lasted for another year. The accusation comprised forty-nine points, plus a number of supplementary charges—for instance, that the accused had failed to shed tears when admonished with texts from Holy Scripture (this "weeping test" was important evidence in witch trials); to which Ma Kepler retorted angrily that she had shed so many tears in her life that she had none left.

The Act of Accusation, read in September, was answered a few weeks later by an Act of Contestation by Kepler and counsel; this was refuted by an Act of Acceptation by the prosecution in December; in May, next year, the defense submitted an Act of Exception and Defense; in August, the prosecution answered with an Act of Deduction and Confutation. The last word was the Act of Conclusion by the defense, one hundred and twenty-eight pages long, and written mostly in Kepler's own hand. After that the case was sent, by order of the duke, to the Faculty of Law at Tuebingen—Kepler's university. The faculty found that Katherine ought to be questioned under torture, but recommended that the procedure should be halted at the stage of *territio,* or questioning under threat of torture.

According to the procedure laid down in such cases, the old woman was led into the torture chamber, confronted with the executioner, the instruments were shown to her, and their action on the body described in detail; then she had a last chance to confess her guilt. The terror of the place was such that a great number of victims broke down and confessed at this stage.[9] The reactions of Ma Kepler were described in the provost's report to the duke as follows:

"Having, in the presence of three members of the court and of the town scribe, tried friendly persuasion on the accused, and having met with contradiction and denial, I led her to the usual place of torture and showed her the executioner and his instruments, and reminded her earnestly of the necessity of telling the truth, and of the great dolor and pain awaiting her. Regardless, however, of all earnest admonitions and reminders, she refused to admit and confess to witchcraft as charged, indicating that one should do with her as one liked, and that even if one artery after another were to be torn from her body, she would have nothing to confess; whereafter she fell on her knees and said a paternoster, and demanded that God should make a sign if she were a witch or a monster or ever had anything to do with witchcraft. She was willing to die, she said; God would reveal the truth after her death, and the injustice and violence done to her; she would leave it all to God, who would not withdraw the Holy Ghost from her, but be her support. . . . Having persisted in her contradiction and denial regarding witchcraft, and having remained steadfast in this position, I led her back to her place of custody."[10]

A week later, Ma Kepler was released, after fourteen months of imprisonment. She could not return to Leonberg, though, because the populace threatened to lynch her. Six months later she died.

It was against this background that Kepler wrote the *Harmony of the World,*[11] in which the third planetary law was given to his engaging contemporaries.

5. "HARMONICE MUNDI"

The work was completed in 1618, three months after the death of his daughter Katherine, and three days after the defenestration of Prague. No irony was intended by the title. Irony he permitted himself only in a footnote (to the sixth chapter of the Fifth Book), where the sounds emitted by the various planets as they hum along their orbits are discussed: "The Earth sings Mi-Fa-Mi, so we can gather even from this that *Mi*sery and *Fa*mine reign on our habitat."

The *Harmony of the World* is a mathematician's Song of Songs "to the chief harmonist of creation"; it is Job's daydream of a perfect universe. If one reads the book concurrently with his letters about the witch trial, his excommunication, the war, and the death of his child, one has the impression of being abruptly transported from one play by his Stratford contemporary to a different one. The letters seem to echo the monologue of King Lear—"Blow, winds, and crack your cheeks! rage! blow! / You cataracts and hurricanoes, spout / Till you have drench'd our steeples, drown'd the cocks! . . . . And thou, all-shaking thunder, / Strike flat the thick rotundity o' the world!" But the book's motto could be, "Here will we sit, and let the sounds of music / Creep in our ears: soft stillness and the night / Become the touches of sweet harmony. . . . There's not the smallest orb which thou behold'st / But in his motion like an angel sings. . . . Such harmony is in immortal souls. . . ."

The *Harmony of the World* is the continuation of the *Cosmic Mystery*, and the climax of his lifelong obsession. What Kepler attempted here is, simply, to bare the ultimate secret of the universe in an all-embracing synthesis of geometry, music, astrology, astronomy, and epistemology. It was the first attempt of this kind since Plato, and it is the last to our day. After Kepler, fragmentation of experience sets in again, science is divorced from religion, religion from art, substance from form, matter from mind.

The work is divided into five books. The first two deal with the concept of harmony in mathematics; the following three with the applications of this concept to music, astrology, and astronomy, in that order.

What exactly does he mean by "harmony"? Certain geometrical proportions that he finds reflected everywhere, the archetypes of universal order, from which the planetary laws, the harmonies of music, the drift of the weather, and the fortunes of man are derived. These geometrical ratios are the *pure* harmonies which guided God in the work of Creation; the *sensory* harmony which we perceive by listening to musical consonances is merely an echo of it. But the inborn instinct in man that makes his soul resonate to music provides him with a clue to the nature of the mathematical harmonies which are at its source. The Pythagoreans had discovered that the octave originates in the ratio 1:2 between the length of the two vibrating strings, the fifth in the ratio of 2:3, the fourth in 3:4, and so on. But they went wrong, says Kepler, when they sought for an explanation of this marvelous fact in occult number lore. The explanation why the ratio 3:5, for instance, gives a concord, but 3:7 a discord, must be sought, not in arithmetical, but in *geometrical* considerations. Let us imagine the string, whose vibrations produce the sound, bent into a circle with its ends joined together. Now a circle can be most gratifyingly divided by inscribing into it symmetrical figures with varying numbers of sides. Thus the side of an inscribed pentagon will divide the circumference into parts which are to the whole circle as 1/5 and 4/5 respectively—both consonant chords.

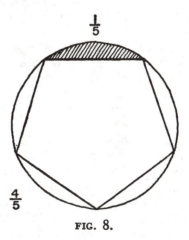

FIG. 8.

But a septagon will produce ratios of 1/7 and 6/7— both discords. Why? The answer, according to Kepler, is: *Because the pentagon can be constructed by compass and ruler, but the septagon not.* Compass and ruler are the only permissible tools in classical geometry. But geometry is the only language that enables man to understand the working of the divine mind. Therefore figures that cannot be constructed by compass and ruler—such as the septagon, the eleven- thirteen- or seventeensided polygons—are somehow unclean, because they defy the intellect. They are *inscibilis*, unknowable,[12] *inefabilis*, unspeakable, *non-entia*, non-existences. "Therein lies the reason," Kepler explains, "why God did not employ the septagon and the other figures of this species to embellish the world."

Thus the pure archetypal harmonies, and their echoes, the musical consonances, are generated by dividing the circle by means of regular polygons; whereas the "unspeakable" polygons produce discordant sounds, and are useless in the scheme of the universe. To the obsession with the five perfect solids was now added the twin obsession with the perfect polygons. The former are three-dimensional bodies inscribed into the sphere; the latter are two-dimensional shapes inscribed into the circle. There is an intimate, mystical connection between the two: the sphere, it will be remembered, is for Kepler the symbol of the Holy Trinity; the two-dimensional plane symbolizes the material world; the intersection of sphere and plane, the circle, pertains to both, and symbolizes the dual nature of man as body and spirit.

But again the facts did not fit the scheme, and had to be explained away by ingenious reasoning. The fifteensided polygon, for instance, can be constructed with ruler and compass, but does not produce a musical consonance. Moreover, the number of construable polygons is infinite, but Kepler only needed seven harmonic relations for his scale (octave, major and minor sixth, fifth, fourth, major and minor third). Also, the harmonies had to be arranged into a hierarchy of varying degrees of "knowability," or perfection. Kepler devoted as much labor to this fantastic enterprise as to the determination of the orbit of Mars. In the end he succeeded, to his own satisfaction, in deriving all his seven harmonies, by certain complicated rules of the game, from his perfect polygons. He had traced back the laws of music to the Supreme Geometer's mind.

In the sections following, Kepler applies his harmonic ratios to every subject under the sun: metaphysics and epistemology; politics, psychology, and physiognomics; architecture and poetry, meteorology and astrology. Then, in the fifth and last book, he returns to cosmology, to complete his vertiginous edifice. The universe he had built in his youth around the five perfect solids had not quite satisfied the observed facts. He now brought the two-dimensional shadow army of polygons to the rescue of the beleaguered solids. The harmonic ratios must somehow be dovetailed in between the solids to fill the gaps and to account for the irregularities.

But how could this be done? How could the harmonies be fitted into the scheme of a universe full of elliptic orbits and non-uniform motions, from which, in fact, all symmetry and harmony seemed to have departed? As usual, Kepler takes the reader into his confidence, and for his benefit recapitulates the process by which he arrived at his solution. At first, he tried to assign the harmonic ratios to the *periods of revolution* of the various planets. He drew a blank. "We conclude that God the Creator did not wish to introduce harmonic proportions into the durations of the planetary years."[18]

Next, he wondered whether the *sizes* or *volumes* of the various planets form a harmonic series. They do not. Third, he tried to fit the greatest and smallest *solar distances* of every planet into a harmonic scale. Again no good. In the fourth place, he tried the ratios between the *extreme velocities* of each planet. Again no good. Next, the variations in the time needed by a planet to cover a *unit length* of its orbit. Still no good. Last, he hit on the idea of transferring the observer's position into the center of the world, and to examine the variations in angular velocity, regardless of distance, *as seen from the sun*. And lo! it worked.

The results were even more gratifying than he had expected. Saturn, for instance, when farthest away from the sun, in its aphelion, moves at the rate of 106 seconds' arc per day; when closest to the sun, and its speed is at maximum, at 135 seconds' arc per day. The ratio between the two extreme velocities is 106 to 135, which differs only by two seconds from 4:5—the major third. With similar, very small deviations (which were all perfectly explained away at the end), the ratio of Jupiter's slowest to its fastest motion is a minor third, Mars' the quint, and so forth. So much for each planet considered by itself. But when he compared the extreme angular velocities of *pairs* of different planets, the results were even more marvelous.

"At the first glance the Sun of Harmony broke in all its clarity through the clouds."[14]

The extreme values yield in fact the intervals of the complete scale. But not enough: if we start with the outermost planet, Saturn, in the aphelion, the scale will be in the major key; if we start with Saturn in the perihelion, it will be in the minor key. Last, if several planets

FIG. 9. *The harmonies of the planets.*
From Harmonices Mundi, *Liber* V (1619).

are simultaneously at the extreme points of their respective orbits, the result, Kepler says, is a motet, where Saturn and Jupiter represent the bass, Mars the tenor, Earth and Venus the contralto, Mercury the soprano. On some occasions, all six can be heard together.

"The heavenly motions are nothing but a continuous song for several voices (perceived by the intellect, not by the ear); a music which, through discordant tensions, through sincopes and cadenzas, as it were (as men employ them in imitation of those natural discords), progresses toward certain predesigned, quasi six-voiced clausulas, and thereby sets landmarks in the immeasurable flow of time. It is, therefore, no longer surprising that man, in imitation of his Creator, has at last discovered the art of figured song, which was unknown to the ancients. Man wanted to reproduce the continuity of cosmic time within a short hour, by an artful symphony for several voices, to obtain a sample test of the delight of the Divine Creator in his works, and to partake of his joy by making music in the imitation of God."[15]

The edifice was complete. Kepler finished the book on May 27, 1618, in one of the most fateful weeks of European history.

"In vain does the God of War growl, snarl, roar, and try to interrupt with bombards, trumpets, and his whole tarantantaran. . . .[16] Let us despise the barbaric neighings which echo through these noble lands, and awaken our understanding and longing for the harmonies."[17]

Out of the murky abyss he soared to heights of orphic ecstasies.

"The thing which dawned on me twenty-five years ago before I had yet discovered the five regu-

lar bodies between the heavenly orbits . . . which
sixteen years ago I proclaimed as the ultimate aim
of all research; which caused me to devote the best
years of my life to astronomical studies, to join
Tycho Brahe and to choose Prague as my residence
—that I have, with the aid of God, who set my
enthusiasm on fire and stirred in me an irrepressible desire, who kept my life and intelligence
alert, and also provided me with the remaining
necessities through the generosity of two emperors
and the estates of my land, Upper Austria—that I
have now, after discharging my astronomical duties
ad satietatum, at long last brought to light. . . .
Having perceived the first glimmer of dawn eighteen months ago, the light of day three months ago,
but only a few days ago the plain sun of a most
wonderful vision—nothing shall now hold me back.
Yes, I give myself up to holy raving. I mockingly
defy all mortals with this open confession: I have
robbed the golden vessels of the Egyptians to make
out of them a tabernacle for my God, far from the
frontiers of Egypt. If you forgive me, I shall rejoice. If you are angry, I shall bear it. Behold, I
have cast the dice, and I am writing a book either
for my contemporaries, or for posterity. It is all the
same to me. It may wait a hundred years for a
reader, since God has also waited six thousand years
for a witness. . . ."[18]

6. THE THIRD LAW

This last quotation is from the preface to the Fifth
Book of the *Harmonice Mundi*, which contains, almost
hidden among the luxuriant growth of fantasy, Kepler's Third Law of planetary motion.

As Kepler proposed it, the Third Law says, "The ratio
which exists between the periodic times of any two

planets is precisely the ratio of 3/2th power of the mean distances, i.e., of the spheres themselves." Put into modern terms, it says that the squares of the periods of revolution of any two planets are as the cubes of their mean distances from the sun.[19] Here is an illustration of it. The earth's period of revolution is one year, and Saturn's is thirty years. The cube root of 1 is 1, and 1 squared is 1, our unit measure of the earth's mean distance. The cube root of 30 is a little more than 3, and 3 squared something over 9, and Saturn's mean distance is, in fact, something more than nine times the earth's. Apologies for the coarse example—it is Kepler's own.[20]

Unlike his First and Second Laws, which he found by that peculiar combination of sleepwalking intuition and wide-awake alertness for clues—a mental process on two levels, which drew mysterious benefits out of his apparent blunderings—the Third Law was the fruit of nothing but patient, dogged trying. When, after endless trials, he had at last hit on the square-to-cube ratio, he of course promptly found a reason why it should be just that and none other; I have said before that Kepler's a priori proofs were often invented a posteriori.

The exact circumstances of the discovery of the Third Law were again faithfully recorded by Kepler.

"On March 8 of this present year 1618, if precise dates are wanted, [the solution] turned up in my head. But I had an unlucky hand and when I tested it by computations I rejected it as false. In the end it came back again to me on May 15, and in a new attack conquered the darkness of my mind; it agreed so perfectly with the data which my seventeen years of labor on Tycho's observations had yielded that I thought at first I was dreaming, or that I had committed a *petitio principi* [begging of the question]. . . ."[21]

He celebrated his new discovery, as he had celebrated

his First Law, with a quotation from Virgil's *Eclogues;* in both, Truth appears in the shape of a teasing hussy who surrenders unexpectedly to her pursuer when he has already given up hope. And in both also, the true solution was rejected by Kepler when it first occurred to him, and was only accepted when it crept in a second time, "through a back door of the mind."

He had been searching for this Third Law—that is to say, for a correlation between a planet's *period* and its *distance*—since his youth. Without such a correlation, the universe would make no sense to him; it would be an arbitrary structure. If the sun had the power to govern the planets' motions, then that motion must *somehow* depend on their distance from the sun; but how? Kepler was the first who saw the problem—quite apart from the fact that he found the answer to it, after twenty-two years of labor. The reason why nobody before him had asked the question is that nobody had thought of cosmological problems in terms of actual physical forces. So long as cosmology remained divorced from physical causation in the mind, *the right question could not occur* in that mind.

The offspring of a new synthesis is not a ready solution, but a healthy problem crying lustily for an answer. And vice versa: a one-sided philosophy—whether it be scholasticism or nineteenth-century mechanism—creates sick problems, of the sort: "What is the sex of the angels?" or "Is man a machine?"

7. THE ULTIMATE PARADOX

The *objective* importance of the Third Law is that it provided the final clue for Newton; hidden away in it is the essence of the law of gravity. But its *subjective* importance to Kepler was that it furthered his chimerical quest—and nothing else. The Law makes its first appearance as "Proposition No. 8" in a chapter characteristically called "The Main Propositions of Astronomy which Are Needed for the Investigation of the Celestial Harmonies." In the same chapter (the only one in the book which deals with astronomy proper) the First Law is merely mentioned in passing, almost shamefacedly, and the Second Law not at all. In its place Kepler once more quoted his faulty inverse-ratio proposition, whose incorrectness he once knew and then forgot. Not the least achievement of Newton was to spot the three Laws in Kepler's writings, hidden away as they were, like forget-me-nots in a tropical flower bed.

To change metaphors once more: the three Laws are the pillars on which the edifice of modern cosmology rests; but to Kepler they meant no more than bricks among other bricks for the construction of his baroque temple, designed by a moonstruck architect. He never realized their real importance. In his earliest book he had remarked that "Copernicus did not know how rich he was"; the same remark applies to Kepler himself.

I have stressed this paradox over and again; now it is time to try to resolve it. First, Kepler's obsession with a cosmos built around the Pythagorean solids and the musical harmonies was not quite as extravagant as it seems to us. It was in keeping with the traditions of Neo-Platonism, with the revival of Pythagoreanism, with the teaching of Paracelsians, Rosicrucians, astrologers, alchemists, cabalists, and hermetists, who were still conspicuously in evidence in the early seventeenth century. When we talk of "the age of Kepler and Galileo," we are likely to forget that they were isolated individuals, a generation ahead of the most enlightened men of their time. If the "harmony of the world" was a fantastic dream, its symbols had been shared by a whole dreaming culture. If it was an *idée fixe*, it was derived from a collective obsession—only more elaborate and precise, enlarged on a grandiose scale, more artful and self-consistent, carried to the ultimate perfection of mathemati-

cal detail. The Keplerian cosmos is the crowning
achievement of a type of cosmic architecture that began
with the Babylonians and ends with Kepler himself.

The paradox, then, is not in the mystic nature of
Kepler's edifice but in the modern architectural elements which it employed, in its combination of incompatible building materials. Dream architects are not
worried about imprecisions of a fraction of a decimal;
they do not spend twenty years with dreary, heartbreak-
ing computations to build their fantasy towers. Only
some forms of insanity show this pedantic method in
madness. In reading certain chapters of the *Harmonice*,
one is indeed reminded of the explosive yet painstak-
ingly elaborate paintings by schizophrenics, which would
pass as legitimate art if painted by a savage or a child,
but are judged by clinical standards if one knows that
they are the work of a middle-aged chartered accountant.
The Keplerian schizophrenia becomes apparent only
when he is judged by the standard of his achievements
in optics, as a pioneer of the differential calculus, the
discoverer of the three Laws. His split mind is revealed
in the manner in which he saw himself in his non-
obsessional moments: as a sober "modern" scientist,
unaffected by any mystic leanings. Thus he writes about
the Scottish Rosicrucian, Robert Fludd:

> "It is obvious that he derives his main pleasure
> from unintelligible charades about the real world,
> whereas my purpose is, on the contrary, to draw the
> obscure facts of nature into the bright light of
> knowledge. His method is the business of alche-
> mists, hermetists, and Paracelsians; mine is the task
> of the mathematician."[22]

These words are printed in *Harmonice Mundi*,
which is buzzing with astrological and Paracelsian ideas.

A second point is equally relevant to the Keplerian
paradox. The main reason why he was unable to realize

how rich he was—that is, to understand the significance
of his own Laws—is a technical one: the inadequacy of
the mathematical tools of his time. Without differential
calculus and/or analytical geometry, the three Laws
show no apparent connection with each other—they are
disjointed bits of information which do not make much
sense. Why should God will the planets to move in ellipses? Why should their speed be governed by the area
swept over by the radius vector, and not by some more
obvious factor? Why should the ratio between distance
and period be mixed up with cubes and squares? Once
you know the inverse square law of gravity and Newton's
mathematical equations, all this becomes beautifully
self-evident. But without the roof which holds them together, Kepler's Laws seem to have no particular *raison
d'être*. Of the first he was almost ashamed: it was a
departure from the circle sacred to the ancients, sacred
even to Galileo and, for different reasons, to himself.
The ellipse had nothing to recommend it in the eyes of
God and man; Kepler betrayed his bad conscience when
he compared it to a cartload of dung which he had to
bring into the system as a price for ridding it of a
vaster amount of dung. The Second Law he regarded
as a mere calculating device, and constantly repudiated
it in favor of a faulty approximation; the Third as a
necessary link in the system of harmonies, and nothing
more. But then, without the notion of gravity and the
method of the calculus, it *could* be nothing more.

Johannes Kepler set out to discover India and found
America. It is an event repeated over and again in the
quest for knowledge. But the result is indifferent to the
motive. A fact, once discovered, leads an existence of its
own, and enters into relations with other facts of which
their discoverers have never dreamed. Apollonius of
Perga discovered the laws of the useless curves which
emerge when a plane intersects a cone at various angles:
these curves proved, centuries later, to represent the

chapter ten

Computing a Bride

Only one circumstance, but a basic one, relieved the
gloom of Kepler's later years: his second marriage, in
1613, to Susanna Reuttinger. He was forty-one, she
twenty-four, the daughter of a cabinetmaker. Susanna's
parents had died while she was a child; she had been
brought up in the household of the Baroness Starhem-
berg. We do not know what position she occupied in
the household, but to judge by the scandalized reactions
of Kepler's correspondents, it must have been a lowly
one—something between a maid and a companion.

Kepler's first marriage had been engineered by his
well-wishers when he was an inexperienced and penniless young teacher. Before his second marriage, friends
and go-betweens again played a prominent part—but
this time Kepler had to choose from among no fewer
than eleven candidates for his hand. In a letter to an
unknown nobleman, which extends to eight printed folio
pages, Kepler has described in meticulous detail the
process of elimination and selection that he followed. It
is a curious document, and among the most revealing
in his voluminous writings. It shows that he solved the
problem of choosing the right wife among the eleven
candidates by much the same method by which he
found the orbit of Mars: he committed a series of mis-

takes which might have proved fatal, but canceled out;
and up to the last moment he failed to realize that he
held the correct solution in his hands.

The letter is dated from Linz, October 23, 1613.[1]

"Though all Christians start a wedding invitation
by solemnly declaring that their marriage is due to
special divine management, I, as a philosopher,
would like to discourse with you, O wisest of men,
in greater detail about this. Was it Divine Providence or my own moral guilt which, for two years
or longer, tore me in so many different directions
and made me consider the possibilities of such different unions? If it was Divine Providence, to what
purpose did it use these various personalities and
events? For there is nothing that I would like to
investigate more thoroughly, and that I more intensely long to know, than this: can I find God,
whom I can almost touch with my hands when I
contemplate the universe, also in my own self? If,
on the other hand, the fault was mine, in what did
it consist? Cupidity, lack of judgment, or ignorance?
And why, on the other hand, was there nobody
among my advisers to approve of my final decision?
Why am I losing their previous esteem or appear
to be losing it?

What could have seemed more reasonable than
that I, as a philosopher, past the peak of virility,
at an age when passion is extinct, the body dried
and softened by nature, should have married a
widow who would look after the household, who
was known to me and my first wife, and unmistak-
ably recommended to me by her? But if so, why
did nothing come of it? . . ."

The reasons why this first project came to nothing
were, among others, that the prospective bride had two

marriageable daughters, that her fortune was in the hands of a trustee; and, as an afterthought:

"also a consideration of health, because, though her body was strong, it was suspect of ill health because of her stinking breath; to this came my dubious reputation in matters of religion. In addition to this, when I met the woman after everything had been settled (I had not seen her for the last six years), there was nothing about her that pleased me. It is therefore sufficiently clear that the matter could not succeed. But why did God permit that I should be occupied with this project which was doomed to failure? Perhaps to prevent my getting involved in other perplexities while my thoughts were on this person? . . . I believe that things like this happen to others too, not only once but often; but the difference is that others do not worry as much as I do, that they forget more easily and get over things quicker than I do; or that they have more self-control and are less credulous than I am. . . . And now for the others.

Together with the mother, her two daughters were also offered to me—under an unfavorable omen, if an offense to probity can be interpreted as such: for the project was presented by the well-wishers of the ladies in a form which was not very proper. The ugliness of this project upset me intensely; yet I began nevertheless to inquire into the conditions. As I thus transferred my interest from widows to virgins, and continued to think of the absent one [the mother] whom, so far, I had not seen, I was captivated by the appearance and pleasant features of the one who was present [the daughter]. Her education had been, as it became sufficiently clear, more splendid than it would be useful to me. She had been brought up in luxury

that was above her station, also she was not of sufficient age to run a household. I decided to submit
the reasons which spoke against the marriage to the
judgment of the mother, who was a wise woman
and loved her daughter. But it would have been
better if I had not done so, because the mother did
not seem to be pleased. This was the second one,
and now I come to the third."

The third was a maiden in Bohemia whom Kepler
found attractive, and who took a liking to his orphaned
children. He left them for a while in her care—"which
was a rash act, for later on I had to fetch them back at
my own expense." She was willing to marry him, but
she had, a year earlier, given her word to another man.

The fourth he would have married gladly, in spite of
her "tall stature and athletic build," if meanwhile the
fifth had not entered the scene. The fifth was Susanna,
his future wife.

"In comparing her to the fourth the advantage
was with the latter as regards the reputation of the
family, earnestness of expression, property, and
dowry; but the fifth had the advantage through her
love, and her promise to be modest, thrifty, diligent, and to love her stepchildren. . . . While I
was waging my long and heavy battle with this
problem, I was waiting for the visit of Frau Helm-
hard, wondering whether she would advise me to
marry the third, who would then carry the day over
the last-mentioned two. Having at last heard what
this woman had to say, I began to decide in favor
of the fourth, annoyed that I had to let the fifth go.
As I was turning this over, and on the point of
making a decision, fate intervened: the fourth got
tired of my hesitations and gave her word to another suitor. Just as I had been previously annoyed
about having to reject the fifth, I was now so much

hurt about the loss of the fourth that the fifth too began to lose her attraction for me. In this case, to be sure, the fault was in my feelings.

Concerning the fifth, there is also the question why, since she was destined for me, God permitted that in the course of one year she should have six more rivals? Was there no other way for my uneasy heart to be content with its fate than by realizing the impossibility of the fulfillment of so many other desires?"

And so to Number Six, who had been recommended to Kepler by his stepdaughter.

"A certain nobility, and some possessions made her desirable; on the other hand, she was not old enough, and I feared the expense of a sumptuous wedding; and her noble rank in itself made her suspect of pride. In addition, I felt pity for the fifth, who had already understood what was afoot and what had been decided. This division in me between willingness and unwillingness had, on the one hand, the advantage that it excused me in the eyes of my advisers, but on the other the disadvantage that I was as pained as if I had been rejected. . . . But in this case too, Divine Providence had meant well, because that woman would not have fitted in at all with my habits and household.

Now, as the fifth ruled, to my joy, alone in my heart, a fact which I also expressed to her in words, suddenly a new rival arose for her, whom I shall call Number Seven—because certain people, whom you know, suspected the humility of the fifth and recommended the noble rank of the seventh. She also had an appearance which deserved to be loved. Again I was prepared to give up the fifth, and to choose the seventh, provided it was true what they said about her. . . ."

But again he prevaricated—"And what else could have been the result but a rejection, which I had quasi-provoked?"

Tongues were wagging all over Linz; to avoid more gossip and ridicule, he now turned his attention to a candidate of common origin "who nevertheless aspired to the nobility. Though her appearance had nothing to recommend her, her mother was a most worthy person." But she was as fickle as he was undecided, and after alternately giving him her word and retracting it on seven subsequent occasions, he again thanked Divine Providence and let her go.

His methods now became more cautious and secretive. When he met Number Nine, who, apart from a lung disease, had much to recommend her, he pretended to be in love with somebody else, hoping that the candidate's reactions might betray her feelings. Her reactions were promptly to tell Mother, who was ready to give her blessing, but Kepler mistakenly thought she had rejected him and then it was too late to put matters right.

The tenth was also of noble rank, of sufficient means and thrifty.

"But her features were most abhorrent, and her shape ugly even for a man of simple tastes. The contrast of our bodies was most conspicuous: I thin, dried-up, and meager; she, short and fat, and coming from a family distinguished by redundant obesity. She was quite unworthy to be compared with the fifth, but this did not revive love for the latter."

The eleventh and last one was again "of noble rank, opulent, and thrifty"; but after waiting four months for an answer, Kepler was told that the maiden was not yet sufficiently grown up.

"Having thus exhausted the counsels of all my

friends, I, at the last moment before my departure for Ratisbon, returned to the fifth, pledged her my word and received hers.

Now you have my commentary on my remark at the beginning of this invitation. You now see how Divine Providence drove me into these perplexities that I may learn to scorn noble rank, wealth, and parentage, of which she has none, and to seek with equanimity other, simpler virtues. . . ."

The letter ends with Kepler's entreating his aristocratic friend to come to the wedding banquet and help him by his presence to brave the adversity of public opinion.

Susanna seemed to have justified Kepler's choice, and lived up to his expectations. There is hardly any later mention of her in his letters, and as far as Kepler's domestic life was concerned, no news is good news. She bore him seven children, of whom three died in infancy.

I have said, at the beginning of this chapter, that Kepler's way of discovering the right wife for himself strangely reminds one of the method of his scientific discoveries. Perhaps, at the end of this matrimonial odyssey, this sounds less farfetched or whimsical. There is the same characteristic split in the personality between, on the one hand, the pathetically eager, Chaplinesque figure who stumbles from one wrong hypothesis to another and from one candidate to the next—oval orbits, egg-shaped orbits, chubby-faced orbits; who proceeds by trial and error, falls into grotesque traps, analyzes with pedantic seriousness each mistake and finds in each a sign of Divine Providence; one can hardly imagine a more painfully humorless performance. But on the other hand, he *did* discover his Laws and *did* make the right choice among the eleven candidates, guided by that sleepwalking intuition which made his waking errors cancel out and always asserted itself at the critical moment. Social rank and financial considerations are topmost in his waking consciousness, yet in the end he married the only candidate who had neither rank, nor money, nor family; and though he anxiously listens to everybody's advice, seems to be easily swayed and without a will of his own, he decides on the person unanimously rejected by all.

It is the same dichotomy that we observed in all his activities and attitudes. In his quarrels with Tycho and constant naggings at him, he displayed embarrassing pettiness. Yet he was curiously devoid of jealousy or lasting resentment. He was proud of his discoveries and often boasted of them (particularly of those which turned out to be worthless), but he had no proprietary feeling about them; he was quite prepared to share the copyright of the three Laws with the *Junker* Tengnagel and, contrary to the habits of the time, gave in all his books most generous credit to others—to Maestlin, Brahe, Gilbert, and Galileo. He even gave credit where none was due, for instance to Fabricius, whom he nearly saddled with the honor of having discovered the elliptic orbits. He freely informed his correspondents of his latest researches and naïvely expected other astronomers to part with their jealously guarded observations; when they refused, as Tycho and his heirs did, he simply pinched the material without a qualm of conscience. He had, in fact, no sense of private property concerning scientific research. Such an attitude is most unusual among scholars in our day; in Kepler's day it seemed quite insane. But it was the most endearing lunacy in his discordant, fantastic self.

chapter eleven

The Last Years

1. "TABULAE RUDOLPHINAE"

Harmonice Mundi was completed in 1618 and published the next year, when Kepler was forty-eight. His pioneering work was done; but in the remaining eleven years of his life he continued to pour out books and pamphlets—annual calendars and ephemerides, a book on comets, another on the new invention of logarithms, and two more major works: the *Epitome Astronomiae Copernicanae* and the *Rudolphine Tables*.

The title of the former is misleading. The *Epitome* is not an abstract of the Copernican system, but a textbook of the Keplerian system. The laws that originally referred to Mars only are here extended to all planets, including the moon and the satellites of Jupiter. The epicycles are all gone, and the solar system emerges in essentially the same shape in which it appears in modern schoolbooks. It was Kepler's most voluminous work and the most important systematic exposition of astronomy since Ptolemy's *Almagest*. The fact that his discoveries are found in it once more side by side with his fantasies does not detract from its value. It is precisely this overlapping of two universes of thought that gives the *Epitome*, as it does to the whole of Kepler's life and work, its unique value to the history of ideas.

To realize how far ahead of his colleagues Kepler was,

in spite of the residue of medievalism in his veins, one must compare the *Epitome* with other contemporary textbooks. None of them had adopted the heliocentric idea, or was to do so for a generation to come. Maestlin published a reprint of his textbook based on Ptolemy in 1624, three years after the *Epitome;* and Galileo's famous *Dialogue on the Great Systems of the World,* published another eight years later, still holds fast to cycles and epicycles as the only conceivable form of heavenly motion.

The second major work of Kepler's late years was his crowning achievement in practical astronomy: the long-awaited Rudolphine Tables, based on Tycho's lifelong labors. Their completion had been delayed for nearly thirty years by Tycho's death, the quarrel with the heirs, and the chaotic conditions created by the war—but basically by Kepler's reluctance against what one might call a Herculean donkey work. Astronomers and navigators, calendar makers and horoscope casters were impatiently waiting for the promised tables, and angry complaints about the delay came from as far as India and the Jesuit missionaries in China. When a Venetian correspondent joined in the chorus, Kepler answered with a cry from the heart:

> "One cannot do everything, as the saying goes. I am unable to work in an orderly manner, to stick to a time schedule and to rules. If I put out something that looks tidy, it has been worked over ten times. Often I am held up for a long time by a computing error committed in haste. But I could pour out an infinity of ideas. . . . I beseech thee, my friends, do not sentence me entirely to the treadmill of mathematical computations, and leave me time for philosophical speculations, which are my only delight."[1]

At last, when he had turned the corner of fifty, he

really settled down to the task at which he had only nibbled since Tycho's death. In December 1623 he triumphantly reported to an English correspondent: *"Video portum"*—"I can see the harbor"; and six months later to a friend: "The Rudolphine Tables, sired by Tycho Brahe, I have carried in me for twenty-two years, as the seed is gradually developed in the mother's womb. Now I am tortured by the labors of birth."[2]

But owing to lack of money and the chaos of the Thirty Years' War, the printing took no less than four years, and consumed half of his remaining energies and life span.

Since the tables were to bear Rudolph's name, Kepler found it fitting that the printing should be financed by payment of the arrears due to him, amounting to 6299 florins. He traveled to Vienna, the new seat of the imperial court, where he had to spend four months to obtain satisfaction. But the satisfaction was of a rather abstract nature. According to the complicated method by which the Crown's financial affairs were transacted, the treasury transferred the debt to the three towns of Nuremberg, Memmingen, and Kempten. Kepler had to travel from town to town—partly on horseback, partly on foot, because of his piles; and to beg, cajole, and threaten, until he finally obtained a total of two thousand florins. He used them to buy the paper for the book, and decided to finance the printing out of his own pocket, "undaunted by any fear for the future sustenance of wife and six children," and though he was forced "to dip into the money held in trust for the children from my first marriage." He had lost a whole year on these travels.

But this was only the beginning of his struggles; the story of the printing of the Rudolphine Tables reminds one of the Ten Egyptian Plagues. To begin with, Linz did not have an adequate printing press for such a major undertaking, so Kepler had to travel again to recruit

skilled printers from other towns. When the work at last
got going, the next plague descended—a familiar one this
time: all Protestants in Linz were ordered either to embrace the Catholic faith, or to leave the town within
six months. Kepler was again exempted, and so was his
Lutheran master printer with his men; but he was requested to hand over to the authorities all books suspect
of heresy. Luckily, the choice of objectionable books
was left to his own judgment (which made him feel "as
if a bitch were asked to surrender one of her litter")
and, thanks to the intervention of the Jesuit Father
Guldin, he was able to keep them all. When the war
was approaching Linz, the authorities asked Kepler's advice how to protect the books of the provincial library
against the danger of fire; he recommended packing
them tightly into wine barrels so that they could easily
be rolled away from the danger spot. Incidentally, notwithstanding his excommunication (now final), Kepler
kept paying visits to his beloved Tuebingen, the Lutheran stronghold, and having a jolly time with old
Maestlin—all which goes to show that the sacred cows of
that bygone age of humanism were still held in respect
during the Thirty Years' War, both in Germany and
Italy, as the case of Galileo shows.

The third plague was the garrisoning of Linz by Bavarian soldiery. Soldiers were billeted everywhere, even
in Kepler's printing shop. This led to a rumor, which
spread through the Republic of Letters, and penetrated
as far as Danzig, that the soldiers had melted down
Kepler's lead type to make bullets and pulped his manuscripts for use as cartridge cases—but luckily this was not
true.

Next, the Lutheran peasantry rose in bloody revolt,
burned down monasteries and castles, occupied the
township of Wels, and laid siege to Linz. The siege
lasted for two months, June–August 1626. There were
the usual epidemics, and the populace was reduced to

living on horseflesh, but Kepler, "by the help of God and the protection of my angels," was preserved from this fate.

"You ask me," he wrote to Father Guldin, "what I did with myself during the long siege. You ought to ask what one could do in the midst of the soldiery. The other houses had only a few soldiers billeted in them. Ours is on the city wall. The whole time the soldiers were on the ramparts, a whole cohort lay in our building. The ears were constantly assailed by the noise of the cannon, the nose by evil fumes, the eye by flames. All doors had to be kept open for the soldiers, who, by their comings and goings, disturbed sleep at night, and work during daytime. I nevertheless considered it a great boon that the head of the Estates had given me rooms with a view over the moats and suburbs in which the fighting took place."[3]

When he did not watch the fighting, Kepler, in his unquiet study, was engaged with his old occupational therapy, the writing of a chronological work.

On June 30, however, the peasants succeeded in setting fire to part of the town. It destroyed seventy houses, and among these was the printing shop. All the sheets that had so far been printed went up in flame; but the angels again intervened and Kepler's manuscript escaped unscathed. This provided him with an occasion for one of his endearing understatements: "It is a strange fate which causes these delays all the time. New incidents keep occurring which are not at all my fault."[4]

Actually, he was not too much aggrieved by the destruction of the printing press, because he had had more than enough of Linz, and was only waiting for a pretext to move elsewhere. He knew of a good press in Ulm, on the upper reaches of the Danube, which belonged to his Swabian homeland, and was less than fifty miles from Tuebingen—that magnetic pole which never lost its attraction. When the siege was lifted and the Emperor's consent obtained, Kepler was able, after fourteen long years, to leave Linz, which he had never liked, and which had never liked him.

But the printer at Ulm turned out to be a disappointment. There were quarrels from the start, and later on threats of a lawsuit. At one point Kepler even left Ulm in a sudden huff to find a better printer—in Tuebingen, of course. He traveled on foot, because he again suffered from boils, which made riding a horse too painful. The time was February, and Kepler was fifty-six. In the village of Blaubeuren, having walked fifteen miles, he turned back and made peace with the printer (whose name was Jonas Saur, meaning sour).

Seven months later, in September 1627, the work was at long last completed. It was just in time for the annual book mart at the Frankfurt fair. Kepler, who had bought the paper, cast some of the type, acted as printer's foreman, and paid for the whole enterprise, now traveled himself to Frankfurt, with part of the first edition of a thousand copies, to arrange for its sale. It was truly a one-man show.

The last of the Egyptian plagues he had to contend with were Tycho's heirs, who now reappeared on the scene. The *Junker* Tengnagel had died five years before, but George de Brahe, the misfired "Tychonides," had continued the guerrilla warfare against Kepler through all these years. He understood nothing of the contents of the work, but he objected to the fact that Kepler's preface occupied more space than his own, and to Kepler's remark that he had improved Tycho's observations, which he regarded as a slur on his father's honor. Since the work could not be published without the heirs' consent, the first two sheets, containing the dedications and prefaces, had to be reprinted twice; as a result, there exist three different versions among the surviving copies of the book.

The *Tabulae Rudolphinae* remained, for more than a century, an indispensable tool for the study of the skies—both planets and fixed stars. The bulk of the work consists of the tables and rules for predicting the positions of the planets, and of Tycho's catalogue of 777 star places, enlarged by Kepler to 1005. There are also refraction tables and logarithms,[5] put for the first time to astronomic uses; and a gazetteer of the towns of the world, their longitudes referred to Tycho's Greenwich—the meridian of Uraniborg-on-Hveen.

The frontispiece, designed by Kepler's hand, shows a Greek temple, under the columns of which five astronomers are engaged in lively dispute: an ancient Babylonian, Hipparchus, Ptolemy, Copernicus, and Tyge de Brahe. In a wall at the base of the temple, under the five immortals' feet, there is a small niche in which Kepler crouches at a rough-hewn working table, mournfully gazing at the onlooker, and to all intents like one of Snow White's Seven Dwarfs. The tablecloth in front of him is covered with numbers, penned by a quill within reach of his hand, indicating the fact that he has no money to buy paper. Over the top of the dome-shaped roof hovers the imperial eagle, dropping gold ducats from its beak, a symbol of imperial largesse. Two of the ducats have landed on Kepler's tablecloth, and two more are falling through the air—a hopeful hint.

2. THE TENSION SNAPS

The last three years of Kepler's life carry haunting echoes of the legend of the Wandering Jew. *"Quis locus eligendus, vastatus an vastandus?"*—"What place shall I choose, one that is destroyed, or one that is going to be destroyed?"[6] He had left Linz forever, and he was without a fixed domicile. Ulm was only a temporary sta-

tion, for the duration of the printing. He was staying in
a house that a friend had put at his disposal, and though
it had been specially altered to accommodate Kepler's
family, he did not have them with him. On the journey
up the Danube from Linz, the river had started to
freeze, and he had to continue by carriage, leaving Susanna and the children midway, at Ratisbon. At least,
that is the explanation he gave in a letter to a correspondent; but he stayed in Ulm nearly ten months, and
did not send for them.

This episode is characteristic of a certain oddness in
his behavior towards the end. It looks as if the heritage
of his vagrant father and uncles was reasserting itself in
his late middle age. His restlessness had found an out-
let in creative achievement; when he finished the Rudolphine Tables, the tension snapped, the current was
cut off, and he seemed to be freewheeling in aimless
circles, driven on by an ever-growing, overriding anxiety.
He was again plagued by rashes and boils; he was afraid
that he would die before the printing of the tables was
finished; and the future was a waste land of famine and
despair.

And yet, in spite of the war, his plight was to a large
part imaginary. He had been offered the most coveted
Chair in Italy, and Lord Bacon's envoy, Sir Henry Wotton, had invited him to England.* Yet he had refused.

"Am I to go overseas where Wotton invites me?
I, a German? I, who love the firm Continent and
who shrink at the idea of an island in narrow bound-
aries of which I feel the dangers in advance?"[7]

After rejecting these tempting offers, he asked in despair his friend Bernegger in Strasburg whether he could
get him a modest lectureship at that university. To attract an audience, he would be willing to cast the horo-

* Kepler had dedicated the *Harmonice Mundi* to James I.

scope of every one of his hearers—because "the threatening attitude of the Emperor, which is apparent in all his words and deeds," left him with hardly any other hope. Bernegger wrote back that his town and university would welcome Kepler with open arms if he were to honor them with his presence, and offered him unlimited personal hospitality in his spacious house with its "very beautiful garden." But Kepler refused, "because he could not afford the expense of the journey"; and when Bernegger tried to cheer him up with the news that a portrait of Kepler was hung on the wall of the university library—"Everybody who visits the library sees it. If only they could see you in person!"—Kepler's reaction was that the protrait "should be removed from that public place, the more so as it has hardly any likeness to myself."[8]

3. WALLENSTEIN

The Emperor's hostility, too, existed in Kepler's imagination only. In December 1627 Kepler left Ulm for Prague—having been almost constantly on the move since the Frankfurt fair—and was received, to his surprise, as *persona grata*. The court had returned to Prague for the coronation of the Emperor's son as king of Bohemia. Everybody was in high spirits: Wallenstein, the new Hannibal, had expelled the Danish invaders from Prussia; he had overrun Holstein, Schleswig, and Jutland, and the enemies of the Empire were everywhere in retreat. Wallenstein himself had arrived in Prague a few weeks before Kepler; he was awarded, in addition to the Duchy of Friedland, which he already held, the Duchy of Sagan in Silesia.

The Emperor's generalissimo and his mathematicus had crossed each other's path before. Wallenstein was addicted to astrology. Twenty years earlier, in Prague, Kepler had been requested, by a go-between, to cast the

nativity of a young nobleman who wished to remain un-
named. Kepler wrote a brilliant character analysis of the
future war leader, who was then twenty-five, which
testifies to his psychological insight—for Kepler had
guessed the identity of his anonymous client.* Sixteen
years later, he was asked, again through a middleman,
to expand the horoscope—which Wallenstein had pro-
fusely annotated on the margin—this time without the
pretense of anonymity. Kepler had again obliged, but
had saved his face with the usual warnings against the
abuses of astrology. This second horoscope, which dates
from 1624, stops at 1634 with the prophecy that March
will bring "dreadful disorders over the land": Wallenstein was murdered on February 25 of that year.†

Thus the ground was prepared for their meeting
amidst the celebrations at Prague. The meeting ended,
after lengthy negotiations, with Kepler's appointment as
Wallenstein's private mathematicus in his newly acquired Duchy of Sagan. The Emperor had no objection,
and Kepler was allowed to retain his title of Imperial
Mathematician for what it was worth—in terms of hard
cash certainly not much; for the Crown's debts to Kepler in arrears of salary and gratuities by now amounted
to 11,817 florins. The Emperor politely informed Wallenstein that he expected the latter to pay this sum—
which Wallenstein, of course, never did.

The deal with Wallenstein concluded, both men left
Prague in May 1628: Wallenstein to lay unsuccessful
siege to Strahlsund, which was the beginning of his
downfall; Kepler to visit his wife and children, who were
still in Ratisbon. He traveled on to Linz, to liquidate
his affairs, then back to Prague, where his family joined

* The name Wallenstein is written in Kepler's secret code
on the original draft of the horoscope, which is still extant.

† But ten years made a round figure at which even a well-paid-
for horoscope could reasonably stop.

him, and in July arrived with them in Sagan. But a considerable part of his possessions, including books and instruments needed for his work, he left behind in storage. It was the halfhearted move of an already broken man, whose behavior became more and more erratic and devious.

Compared to Sagan, Linz had been paradise.

"I am a guest and a stranger here, almost completely unknown, and I hardly understand the dialect of the locals, who in turn consider me a barbarian. . . .[9]

I feel confined by loneliness, far away from the great cities of the Empire; where letters come and go slowly, and at heavy expense. Add to this the agitations of the [counter-] Reformation which, though I am not personally hit, did not leave me untouched. Sad examples are before me, or before my mind's eye, of acquaintances, friends, people of my immediate neighborhood being ruined, and conversation with the terror-stricken is cut off by fear. . . .

A little prophetess of eleven in Kottbuss, which is between here and Frankfort-on-the-Oder, threatens with the end of the world. Her age, her childish ignorance, and her enormous audiences make people believe in her."[10]

It was the same story as in Gratz and Linz: the people were compelled to become Catholics or to leave the country. They were not even allowed to follow a Lutheran hearse to the cemetery. The privileged position that Kepler enjoyed only intensified his loneliness. He was a prisoner of constant, nagging anxieties about matters large and small.

"It seems to me that there is disaster in the air. My agent Eckebrecht in Nuremburg, who handles all my affairs, has not written for two months. . . .

I am worried about everything, about my account in Linz, about the distribution of the tables, about the nautical chart for which I have given a hundred and twenty florins to my agent, about my daughter, about you, about the friends in Ulm."[11]

There was, of course, no printing press in Sagan, so he set out again on travels to procure type, machinery, and printers. This took nearly eighteen months out of the, altogether, two years, the last of his life, that he spent in Sagan.

"Amidst the collapse of towns, provinces, and countries, of old and new generations, in the fear of barbaric raids, of the violent destruction of hearth and home, I see myself obliged, a disciple of Mars though not a youthful one, to hire printers without betraying my fear. With the help of God I shall indeed bring this work to an end, in a soldierly fashion, giving my orders with bold defiance and leaving the worry about my funeral to the morrow."[12]

4. LUNAR NIGHTMARE

When the new press was installed in December 1629, in Kepler's own lodgings, he embarked (with his assistant, Bartsch, whom Kepler had bullied into marrying his daughter Susanna) on a remunerative enterprise: the publication of ephemerides* for the years 1629–36. Ever since the *Tabulae Rudolphinae* had come out, astronomers all over Europe were competing to publish ephemerides, and Kepler was anxious "to join the race" as he said, on the race track that he had built. But in between, he also began the printing of an old, favorite brain child of his: *Somnium*—a dream of a journey to the

* "Ephemerides" provide detailed information about the motions of the planets for a given year, whereas "tables" give only the broad outlines on which the calculations are based.

moon. He had written it some twenty years before, and from time to time had added notes to it, until these had far outgrown the original text.

The *Somnium* remained a fragment; Kepler died before he finished it, and it was only published posthumously, in 1634. It is the first work of science fiction in the modern sense—as opposed to the conventional type of fantasy utopias from Lucian to Campanella. Its influence on later authors of interplanetary journeys was considerable—from John Wilkins' *Discovery of a New World* and Henry More right down to Samuel Butler, Jules Verne, and H. G. Wells.[13]

Somnium starts with a prelude full of autobiographical allusions. The boy Duracotus lived with his mother Fiolxhilda on Iceland, "which the ancients called Thule".* The father had been a fisherman, who had died at the age of one hundred and fifty when the boy was only three. Fiolxhilda sold herbs in little bags of ramskin to the seamen, and conversed with demons. At fourteen, the boy curiously opened one of the little bags, whereupon his mother, in a fit of temper, sold him to a seafaring captain. The captain left him on the Isle of Hveen, where for the next five years Duracotus studied the science of astronomy under Tycho de Brahe. When he returned home, his repentant mother, as a treat, conjured up one of the friendly demons from Lavania†—the moon —in whose company selected mortals might travel to that planet. "After completing certain ceremonies, my mother, commanding silence with her outstretched hand, sat down beside me. No sooner had we, as arranged, covered our heads with a cloth, when a hoarse,

* Kepler chose the name Duracotus because it sounded Scottish, "and Scotland lies on the Islandic ocean"; "Fiolx" was the name for Iceland that he saw on an old map.

† From *Lavanah*—the Hebrew name of the moon. (Lavan = white.)

supernatural voice began to whisper, in the Icelandic language, as follows. . . ."

Thus ends the prelude. The journey itself, the demon explains, is only possible during an eclipse of the moon, and must therefore be completed in four hours. The traveler is propelled by the spirits, but he is subject to the laws of physics; it is at this point that science takes over from fantasy.

"The intial shock [of acceleration] is the worst part of it, for he is thrown upward as if by an explosion of gunpowder. . . . Therefore he must be dazed by opiates* beforehand; his limbs must be carefully protected so that they are not torn from him and the recoil is spread over all parts of his body. Then he will meet new difficulties: immense cold and inhibited respiration. . . . When the first part of the journey is completed, it becomes easier, because on such a long journey the body no doubt escapes the magnetic force of the earth and enters that of the moon, so that the latter gets the upper hand. At this point we set the travelers free and leave them to their own devices: like spiders they will stretch out and contract, and propel themselves forward by their own force—for, as the magnetic forces of the earth and moon both attract the body and hold it suspended, the effect is as if neither of them was attracting it—so that in the end its mass will by itself turn toward the moon."

In the *Astronomia Nova* Kepler had come so close to the concept of universal gravity that one had to assume the existence of a psychological blockage which made him reject it. In the passage just quoted he not only takes it for granted, but, with truly astonishing insight,

* It has recently been suggested that space travelers should be anesthetized during the initial acceleration.

postulates the existence of "zones of zero gravity"—that nightmare of science fiction. Later on in the *Somnium* he took a further step in the same direction by assuming that there are spring tides on the moon, due to the joint attraction of sun and earth.

The journey completed, Kepler proceeds to describe conditions on the moon. A lunar day, from sunrise to sunset, lasts approximately a fortnight, and so does a moon night—for the moon takes a month to turn once around its axis, the same time it takes to complete a revolution around the earth. As a result, it always turns the same face to the earth, which the moon creatures call their "volva" (from *revolvere*, to turn). This face of the moon they call the Subvolvan half; the other is the Prevolvan half. Common to both halves is that their year consists of twelve days and nights, and the resulting dreadful contrasts of temperature—scorching days, frozen nights. Common to both are also the queer motions of the starry sky—the sun and planets scuttle incessantly back and forth, a result of the moon's gyrations round the volva. This "lunatic" astronomy—in the legitimate double meaning of the word—which Kepler develops with his usual precision, is sheer delight; nobody before (nor since, as far as I know) had attempted such a thing. But when it comes to conditions on the moon itself, the picture becomes grim.

The Prevolvans are the worst off. Their long nights are not made tolerable by the presence of the huge volva, as on the other hemisphere, for the Prevolvans of course never see the earth. Their nights are "bristling with ice and snow under the raging, icy winds." The day that follows is no better: for a fortnight the sun never leaves the sky, heating the air to a temperature "fifteen times hotter than our Africa."

The mountains of Lavania are much higher than those on earth; so are the plants and the creatures that inhabit it. "Growth is rapid; everything is short-lived be-

cause it develops to such an enormous bodily mass. . . .
Growth and decay takes place in a single day." The
creatures are mostly like gigantic serpents. "The Prevolvans have no fixed and safe habitations; they traverse in
hordes, in a single day, the whole of their world, following the receding waters either on legs that are longer
than those of our camels, or on wings, or in ships." Some
are divers, and breathe very slowly, so that they can take
refuge from the scorching sun at the bottom of the deep
waters. "Those that remain on the surface are boiled by
the midday sun and serve as nourishment for the approaching nomadic hordes. . . . Others, who cannot live
without breathing, retreat into caves which are supplied
with water by narrow canals so that the water may
gradually cool on its long way and they may drink it;
but when the night approaches, they go out for prey."
Their skin is spongy and porous; but when a creature
is taken unaware by the heat of the day, the skin becomes hard and scorched, and falls off in the evening.
And yet they have a strange love for basking in the sun
at noon—but only close to their crevices, to be able to
make a swift and safe retreat. . . .

In a short appendix, the Subvolvans are allowed cities
surrounded by circular walls—the craters of the moon;
but Kepler is only interested in the engineering problems of their construction. The book ends with Duracotus being awakened by a cloudburst from his dream—
or rather, from his nightmare of prehistoric giant rep-
tiles, of which Kepler had, of course, no knowledge whatsoever. No wonder that Henry More was inspired by the
Somnium to a poem called *Insomnium Philosophicum*.
But more amusing is Samuel Butler's paraphrase on
Kepler in "The Elephant in the Moon."

> *Quoth he—Th' Inhabitants of the Moon,*
> *Who when the Sun shines hot at Noon,*
> *Do live in Cellars underground*

The Last Years

Of eight Miles deep and eighty around
(In which at once they fortify
Against the Sun and th' Enemy)
Because their People's civiler
Than those rude Peasants, that are found
To live upon the upper Ground,
Call'd Privolvans, with whom they are
Perpetually at open War.

Although most of the *Somnium* was written much
earlier, one readily understands why it was the last book
on which he worked, and which he wished to see in
print. All the dragons which had beset his life—from the
witch Fiolxhilda and her vanished husband, down to the
poor reptilian creatures in perpetual flight, shedding
their diseased skin, and yet so anxious to bask under an
inhuman sun—they are all there, projected into a cosmic
scenery of scientific precision and rare, original beauty.
All Kepler's work, and all his discoveries, were acts of
catharsis; it was only fitting that the last one should end
with a fantastic flourish.

5. THE END

Wallenstein could not have cared less what Kepler
did. The arrangement had been a mutual disappointment from the beginning. Unlike the aristocratic dilet-
tantes who had patronized Tycho, Galileo, and Kepler
himself in the past, General Wallenstein took no genuine interest in science. He drew a certain snob satisfaction from having a man of European renown as his court
mathematicus, but what he really wanted from Kepler
was astrological advice regarding the political and military decisions he had to make. Kepler's answers to such
concrete questions were always elusive—owing to his honesty, or caution, or both. Wallenstein used Kepler
mainly to obtain exact data on the planetary motions,

which he then sent on to his more willing astrologers—
like the notorious Seni—as a basis for their auguries.
Kepler himself rarely spoke about his personal contacts
with Wallenstein. Though he once calls him "a second
Hercules,"[14] his feelings were more honestly reflected in
one of his last letters.

> "I have returned recently from Gitschin [Wal-
> lenstein's residence], where my patron kept me in
> attendance for three weeks—it meant a considerable
> waste of time for both of us."[15]

Three months later, the pressure of Wallenstein's
rivals induced the Emperor to dismiss his generalissimo.
It was only a temporary setback in Wallenstein's dramatic career, but Kepler believed that it was the end.
Once again, and now for the last time, he took to the
roads.

In October, he departed from Sagan. He left his family behind, but packed up cartloads of books and documents, which were dispatched ahead to Leipzig. His son-
in-law wrote later on: "Kepler left Sagan unexpectedly,
and his condition was such that his widow, his children
and friends expected to see the Last Judgment sooner
than his return."[16]

His purpose was to look out for another job, and to
try to obtain some of the money owed to him by
the Emperor and by the Austrian Estates. In his self-
analysis, thirty-five years earlier, he had written that his
constant worrying about money "was not prompted by
the desire for riches, but by fear of poverty." This was
still essentially true. He had money deposits in various
places, but he was unable to recover even the interests
due to him. When he set out on that last journey across
half of wartorn Europe he took all the cash he had with
him, leaving Susanna and the children penniless. Even
so, he had to borrow fifty florins from a merchant in
Leipzig, where he stopped on the first lap of his journey.

He seems to have had one of his curious premonitions. All his life he had been in the habit of casting horoscopes for his birthdays. The horoscopes for the years preceding and following his sixtieth show merely the positions of the planets, without comment. The sixtieth, his last, is an exception; he noted on it that the positions of the planets were almost the same as at his birth.

His last letter is dated from Leipzig, October 31, and addressed to friend Bernegger in Strasburg. He had remembered Bernegger's earlier invitation, and suddenly decided to accept it; but he seems to have forgotten it again a moment later, for in the remainder of the letter he talks of his traveling plans without any reference to it.

"Your hospitality I gladly accept. May God preserve you, and take pity on the misery of my country. In the present general insecurity one ought not to refuse any offer of shelter, however distant its location. . . . Farewell to you, your wife and children. Hold fast, with me, to our only anchor, the Church, pray to God for it and for me."[17]

From Leipzig he rode on, on a miserable old horse to Nuremberg, where he visited a printer. Then on to Ratisbon, where the Diet was sitting, in full pomp, presided over by the Emperor who owed him twelve thousand florins.

He arrived in Ratisbon on November 2. Three days later he took to bed with a fever. An eyewitness reported that "he did not talk, but pointed his index finger now at his head, now at the sky above him."[18] Another witness, the Lutheran preacher, Jacob Fischer, wrote in a letter to a friend:[19]

"During the recent session of the Diet, our Kepler arrived in this town on an old jade (which he subsequently sold for two florins). He was only three days here when he was taken ill with a feverish ailment. At first he thought that he was suffering from *sacer ignis,* or fever pustules, and paid no attention to it. When his feverish condition worsened, they bled him, without any result. Soon his mind became clouded with ever-rising fever. He did not talk like one in possession of his faculties. Several preachers visited him and comforted him with the living waters of their sympathy.[20] In his last agony, as he gave up his ghost to God, a Protestant clergyman of Ratisbon, Sigismund Christopher Donavarus, a relative of mine, consoled him in a manly way, as behoves a servant of God. This happened on November 15, 1630. On the 19th he was buried in the cemetery of St. Peter, outside the town."

The cemetery was destroyed in the Thirty Years' War, and Kepler's bones were scattered; but the epitaph which he wrote for himself is preserved.

Mensus eram coelos, nunc terrae metior umbras
Mens coelestis erat, corporis umbra iacet.

I measured the skies, now the shadows I measure
Sky-bound was the mind, earth-bound the body rests.

There is also a paragraph in one of his late letters which lingers on in memory; it is dated:

"Sagan in Silesia, in my own printing press, November 6, 1629:
When the storm rages and the state is threatened by shipwreck, we can do nothing more noble than to lower the anchor of our peaceful studies into the ground of eternity."[21]

Notes

Joannis Kepleri Astronomi Opera Omnia, ed. Ch. Frisch, 8 Vol., Frankofurti et Erlangae, 1858–1871.

A modern collected edition of Kepler's work and correspondence, *Johannes Kepler, Gesammelte Werke,* ed. by W. v. Dyck✠ and Max Caspar, in collaboration with Franz Hammer, was begun in 1938. Up to date (March 1958), Volumes I to VII, IX, XIII to XVII are available. The texts are in the original Latin and medieval German.

The only serious modern work of biography is Max Caspar's *Johannes Kepler,* Stuttgart, 1948.

Abbreviations

    O.O. —*Opera Omnia.*
    G.W.—*Gesammelte Werke.*
    Ca. —Caspar's Biography.

Chapter One THE YOUNG KEPLER

1 O.O., Vol. VIII, p. 670 *seq.,* henceforth referred to as "Horoscope."

2 In 1945, a French unit was advancing on the town and started shelling it in the mistaken belief that the retreating German army had left a rear guard between its walls. At the critical moment a French officer—whose name was given to me as Colonel de Chastigny—arrived at the scene,

identified it as Kepler's birthplace, stopped the firing, and saved Weil from destruction.

3 "One of my ancestors, Heinrich, and his brother, Friedrich, were knighted . . . in 1430, by the Emperor [Sigismond] on the bridge over the Tiber in Rome." (Letter from Kepler to Vincento Bianchi, February 17, 1619; G.W., Vol. XVII, p. 321.) The Patent of Nobility is still extant, but the two Keplers knighted in 1430 were called Friedrich and Konrad, not Friedrich and Heinrich.

4 "Horoscope."

5 Ibid.

6 Ibid.

6a Kretschmer, *The Psychology of Men of Genius* (trans. by R. B. Cattell), London, 1931.

7 I.e., placed very close to the sun.

8 O.O., Vol. V, p. 476 *seq.*, henceforth referred to as "Memoir."

9 "Memoir." Cf. also letter to Herwart von Hohenburg. 9/10.4.1599, G.W., Vol. XIII, p. 305 ff.

10 "Horoscope."

11 *Johannes Kepler in seinen Briefen*, ed. by Caspar and v. Dyck, Munich and Berlin, 1930, Vol. I, p. 26.

12 "Memoir."

13 G.W., Vol. XIII, p. 19 f.

14 *Tertius interveniens*, G.W., Vol. IV, p. 145 *seq.*

15 *De Stella Nova in pede Serpentarii*, G.W., Vol. I, p. 147 *seq.*

16 *Tertius interveniens.*

17 *De Stella Nova*, Cap. 28.

18 *Antwort auf Röslini Diskurs*, G.W., Vol. IV, p. 99 *seq.*

19 Ca., p. 108.

20 *Tertius interveniens.*

21 "Memoir."

21a *Antwort auf Röslini Diskurs*, p. 127.

22 *Tertius interveniens.*

23 To Herwart, G.W., Vol. XIII, p. 305 ff.

Notes

Chapter Two THE "COSMIC MYSTERY"

1 *Mysterium Cosmographicum* (G.W., Vol. I), preface to the reader.
2 Ibid., loc. cit.
3 Loc. cit.
4 Particularly striking is Kepler's advanced relativistic position in the first chapter of the *Mysterium*. For "metaphysical and physical" reasons, he says, the sun must be in the center of the world, but this is not necessary for a formally correct description of the facts. Concerning the Ptolemaic and Copernican views of the apparent motion of the fixed stars, he says: "It is sufficient that both should say [what both really say] that this phenomenon is derived from a contrasting motion between earth and sky." Regarding the annual revolution, he says that the universe of Tycho (where five planets revolve round the sun and the sun revolves round the earth) is pragmatically as legitimate as the Copernican. "Indeed, the proposition 'the sun rests in the center' is too narrow, goes too far. It is sufficient to postulate more generally: 'the sun is the center of the five planets.'"
5 In England, the significance of Copernicus was recognized earlier than on the Continent, mainly thanks to two works: first, Thomas Digges' *A Perfit Description of the Caelestiall Orbes according to the most aunciente doctrine of the Phythagoreans, latelye reuiued by Copernicus and by Geometricall Demonstrations approued*, which he added, in 1576, to a new edition of his father's, Leonard Digges', *Prognostication euerlasting*; and, second, Giordano Bruno's *La cena de le ceneri*, which Bruno wrote during his English sojourn, and which was first published by Charlewood in London in 1584.
6 Cap. 13.
7 By inscribing Mercury's sphere, not into the faces of the octahedron, as it ought to be done, but into the square formed by the four median edges. Cap. 13, Note 4.
8 Cap. 15.
9 Cap. 18.

10 Ibid., Note 8.
11 Cap. 20.
12 Ibid., Notes 2 and 3.
13 The law resulting from this first attempt was: $R_1 : R_2 = P_1 : \dfrac{P_1 + P_2}{2}$, where P_1, P_2 are the periods, R_1, R_2 the mean solar distances of two planets. The correct law (Kepler's Third Law) is, of course: $R_1 : R_2 = P_1^3 : P_2^3$.
14 Cap. 21.
15 Ibid., Note 7.
16 Ca., p. 78.
17 *Mysterium Cosmographicum*, Dedication of the 2nd Edition.
18 *Astronomia Nova*, summary of Cap. 45.
19 Letter to Maestlin, 3.10.1595. G.W., Vol. XIII, p. 33 ff.
20 *Tertius interveniens.*
21 *Harmonice Mundi*, Lib. IV, Cap. I. G.W., Vol. VI.
22 *Mysterium Cosmographicum*, Cap. 21, Notes 8 and 11.
23 It is curious to note that no authority writing on Kepler seems to have noticed this stubborn omission of the word "ellipse"; perhaps because historians of science recoil from the irrationality of their heroes, as Kepler himself recoiled from the apparent irrationality of the elliptic orbits which he discovered.
24 Burtt, *The Metaphysical Foundations of Modern Physical Science* (rev. ed.), London, 1932, p. 203. Burtt is a notable exception from the attitude referred to in the previous note.
25 *Tertius interveniens.*
26 *Mysterium Cosmographicum*, Preface to the Reader.
26a Ibid., Note 8.

Chapter Three GROWING PAINS

1 Letter to Friedrich, Duke of Wuerttemberg, 27.2.1596. G.W., Vol. XIII, p. 50 ff.
2 G.W., Vol. XIII, p. 162 ff.
3 Letter to Maestlin, 11.6.1598. G.W., Vol. XIII, p. 218 *seq.*

4 "Horoscope." Cf. also letter to Maestlin, 10.2.1597, G.W., Vol. XIII, p. 104 *seq*.

5 Letter to Maestlin, 9.4.1597, G.W., Vol. XIII, p. 113 *seq*.

6 Letter to Herwart, 9/10.4.1599, G.W., Vol. XIII, p. 305 *seq*.

7 Letter to an anonymous woman, *c*. 1612, G.W., Vol. XVII, p. 39 *seq*.

8 Ibid.

9 E. Reicke, *Der Gelehrte, Monographien zur deutschen Kulturgeschichte*, Vol. VII, Leipzig, 1900, p. 120.

10 G.W., Vol. XIII, p. 84 f.

11 G.W., Vol. XIII, p. 207.

12 Letter to Herwart, 16.12.1598, G.W., Vol. XIII, p. 264 *seq*.

13 From the failure of his efforts, Kepler concluded that the parallax of the Polar Star must be smaller than 8', "because my instrument does not allow me to measure angles smaller than this. Hence the semi-diameter of the earth's orbit must be smaller than 1/500 of the semi-diameter of the sphere of the fixed stars." (Letter to Herwart, G.W., Vol. XIII, p. 267 f.).

Copernicus assumed the mean distance of the earth from the sun = 1142 earth radii (*De Revolutionibus*, Lib. IV, Cap. 21). In round figures, the radius of the earth's orbit thus amounts to 1200 × 4000 = 4.8 million miles; and the minimum radius of the universe to 4.8 × 500 = 2400 million miles. Later on, however, in the *Epitome*, he enlarged the radius of the universe to sixty million earth radii, i.e., 24.10^{10} miles. He arrived at this figure by assuming that the orbital radius of Saturn was the geometrical mean between the radius of the sun and the radius of the sphere of the fixed stars; and the radius of the sun to be fifteen times the radius of the earth. (*Epitome*, IV, 1, O.O., Vol. VI, p. 332.)

14 Letter to Herwart, 16.12.1598, loc. cit. Kepler himself never accepted infinity. He believed that the fixed stars were all placed at almost exactly the same distance from the sun, so that their "sphere" (which, of course, he did not regard as real) was only "two German miles" in thickness. (*Epitome*, IV, 1, O.O., Vol. VI, p. 334.)

15 Letter to Maestlin, 16/26.2.1599, G.W., Vol. XIII, p. 289 *seq.*

16 Ibid.

17 Letter to Maestlin, 8.12.1598, G.W., Vol. XIII, p. 249 *seq.*

18 12.9.1597, G.W., Vol. XIII, p. 131 *seq.*

19 Letter to Herwart, 16.12.1598, G.W., Vol. XIII, p. 264 *seq.*

20 Letter to Maestlin, 29.8.1599, G.W., Vol. XIV, p. 43 *seq.*

21 Letter to Maestlin, 22.11.1599, G.W., Vol. XIV, p. 86 *seq.*

22 Maestlin to Kepler, 25.1.1600, G.W., Vol. XIV, p. 105 *seq.*

Chapter Four TYCHO DE BRAHE

1 J. L. E. Dreyer, *Tycho Brahe*, Edinburgh, 1890, p. 27. Dreyer's is the modern standard biography of Tycho. He also edited Tycho's *Opera Omnia.*

2 Loc. cit.

3 Op. cit., p. 14.

4 To be precise, he used two threads, passing through two pairs of stars, and intersecting in the nova.

5 Op. cit., p. 86 f.

6 *An Itinerary written by Fynes Morison, etc.*, London, 1617, fol., p. 60, quoted by Dreyer, p. 89.

7 Dreyer, op. cit., p. 105.

8 Ibid, p. 262 n.

8a His other principal achievements were: improved approximations of the sun's and moon's orbits; discovery of the "moon's equation" (independently of Kepler); demolition of the Copernican belief in a periodic inequality in the precession of the equinoxes.

9 Ibid., p. 261.

10 Ibid., p. 249 f.

11 Ibid., p. 279.

12 *Nicolai Raimari Ursi Dithmarsi Fundamentum astronomicum*, Strasburg, 1588.

13 The only differences between the system of Ursus and the

Notes

Tychonic system were that in the former the daily rotation was attributed to the earth, in the latter to the fixed stars; and that different orbits were attributed to Mars.

14 To Ursus, 15.11.1595, G.W., Vol. XIII, p. 48 f.
15 *Nicolai Raimari Ursi Dithmarsi de astronomicis Hypothesibus*, etc., Prague, 1597.
16 To Tycho, 13.12.1597, G.W., Vol. XIII, p. 154.
17 Tycho to Kepler, 1.4.1598, G.W., Vol. XIII, p. 197 *seq.*
18 21.4.1598, G.W., Vol. XIII, p. 204 f.
19 19.2.1599, G.W., Vol. XIII, p. 286 f.
20 The passage runs: "Some doctor stopped on his return journey from Italy in Gratz and showed me a book of his [Ursus'] which I hurriedly read in the three days for which I was permitted to keep it. I found in it . . . certain golden rules which, as I remembered, Maestlin had frequently used in Tuebingen, and also the science of the *sine* and of the computation of triangles—subjects which, though generally known, were new to me . . . for afterward I found in Euclid and Regiomontanus most of what I had ascribed to Ursus."
21 G.W., Vol. XIV, p. 89 *seq.*

Chapter Five TYCHO AND KEPLER

1 Dreyer, op. cit., p. 279.
2 Letter to Herwart, 12.7.1600, G.W., Vol. XIV, p. 128 *seq.*
3 Ca., p. 117.
4 To Herwart, 12.7.1600, G.W., Vol. XIV, p. 128 *seq.*
5 Ca., p. 119.
6 Tycho to Jessenius, 8.4.1600, G.W., Vol. XIV, p. 112 *seq.*
7 April 1600, G.W., Vol. XIV, p. 114 *seq.*
7a He had signed, however, a written undertaking to keep all information he obtained from Tycho "in highest secrecy," i.e., he could publish nothing without Tycho's consent.
8 9.9.1600, G.W., Vol. XIV, p. 150 *seq.*
9 9.10.1600, G.W., Vol. XIV, p. 155 *seq.*
10 28.8.1600, G.W., Vol. XIV, p. 145 *seq.*

11 F. Morison, op. cit.
12 Dreyer, op. cit., p. 386 f.
13 Quoted by Kepler in *Astronomia Nova*, I, Cap. 6.

Chapter Six   THE GIVING OF THE LAWS

1 ASTRONOMIA NOVA ΑΙΤΙΟΛΟΓΗΤΟΣ, *sev*
  *PHYSICA COELESTIS, tradita commentariis DE
  MOTIBUS STELLÆ MARTIS, Ex observationibus,
  G. V. TYCHONIS BRAHE.*
1a *Astronomia Nova*, G.W., Vol. III, preamble to the table
  of contents.
2 Ibid., II, Cap. 7.
3 Ibid., Dedication.
4 "It is inconceivable that a non-material force should be
  present in a non-body and should move through space and
  time," ibid., I, Cap. 2.
5 Ibid., II, Cap. 14.
6 Ibid., II, Cap. 14.
7 At a later stage, however, he reverted to the Ptolemaic position.
8 Altogether, Tycho had observed ten oppositions, and Kepler himself two (1602 and 1604). The Tychonic data
  which he used were those for 1587, 1591, 1593, and 1595.
8a Letter to Herwart, 12.7.1600, G.W., Vol. XIV, p. 132 f.
9 *Astronomia Nova*, II, Cap. 18.
10 Ibid., II, Cap. 19.
11 *Science and the Modern World* (reprint), Cambridge,
   1953, p. 3.
12 *Astronomia Nova*, II, Cap. 20, III, Cap. 24.
13 Ibid., III, Cap. 22.
14 Loc. cit.
15 The observer on Mars went into action each time Mars
   returned to a given position in its orbit, i.e., had the same
   heliocentric longitude. Since the sidereal period of Mars
   was known, the times when this occurred could be determined, and the different positions the earth occupied at
   these times could also be determined. The method yielded
   a series of triangles, Mars–Sun–Earth: MSE_1, MSE_2,

MSE_3, where the angles at S and E were known (from Tycho's data and/or from Kepler's previously established method of approximation). These yielded the ratios SE_1/SM, SE_2/SM, SE_3/SM; and it was now a simple problem in geometry to determine the earth's orbit, (still assumed to be circular), its eccentricity, and the position of the *punctum equans*. The same method enabled him later on to determine the relative Mars–Sun distances for any observed geocentric longitude of Mars.

16 At the beginning of III, Cap. 33.

17 Table of contents, summary of Cap. 32.

18 "At other places [not in the vicinity of aphelion and perihelion] there is a very small deviation. . . ." The passage implies that the deviation is negligible. This is true of the earth's orbit, because of its small eccentricity, but not at all true of Mars, with its large eccentricity.

19 That Descartes derived his theory of vortices from Kepler is probable, but unproven.

20 *Astronomia Nova*, III, Cap. 40.

21 Loc. cit.

22 Loc. cit.

23 To sum up, the three incorrect assumptions are: (a) that the planet's velocity varies in inverse ratio with its distance from the sun; (b) the circular orbit; (c) that the sum of eccentric radii vectors equals the area. The erroneous *physical* hypotheses played only an indirect part in the process.

24 Letter to Longomontanus, 1605, G.W., Vol. XV, p. 134 *seq.*

25 *Astronomia Nova*, IV, Cap. 45.

26 Loc. cit.

27 Letter to D. Fabricius, 18.12.1604, G.W., Vol. XV, p. 78 *seq.*

28 Letter to D. Fabricius, 4.7.1603, G.W., Vol. XIV, p. 409 *seq.*

29 Letter to D. Fabricius, 18.12.1604.

30 *Astronomia Nova*, IV, Cap. 55.

31 Ibid., IV, Cap. 56.

32 Ibid., IV, Cap. 58.

33 1605, G.W., Vol. XV, p. 134 *seq.*

34 *Mysterium Cosmographicum*, Cap. 18.
35 Cf. *Insight and Outlook*, London and New York, 1949.
35a *Astronomia Nova*, introduction.
36 Delambre, *Histoire de l'Astronomie Moderne*, Paris, 1891, Vol. I, p. 394.
37 Third letter to Bentley, *Opera Omnia*, London, 1779–85, IV, 380. Quoted by Burtt, op. cit., p. 265 f.
38 Thus, for instance, in Galileo's *Dialogue on the Great World Systems*, it is Simplicius, the naïve Aristotelian, who says: "The cause [why bodies fall] is manifest, and everyone knows that it is gravity"; but he is promptly rebuked with: "You are out, Simplicius; you say that everyone knows that it is *called* gravity, and I do not question you about the name but about the essence of the thing. Of this you know not a tittle more than you know the essence of the mover of the stars in gyration." (Salusbury trans., ed. de Santillana, Chicago, 1953, p. 250.)
39 10.2.1605, G.W., Vol. XV, p. 145 *seq.*
40 *Astronomia Nova*, III, Cap. 33.
41 Ibid., III, Cap. 38.
42 Ibid., I, Cap. 6.
43 Max Caspar's introduction to his German translation of the *Astronomia Nova*, Munich and Berlin, 1929, p. 54.

Chapter Seven KEPLER DEPRESSED

1 Letter to Heydon, October 1605, G.W., Vol. XV, p. 231 *seq.*
2 Letter to D. Fabricius, 1.10.1602, G.W., Vol. XIV, p. 263 *seq.*
2a Letter to D. Fabricius, February 1604, G.W., Vol. XV, p. 17 *seq.*
3 "Greetings to the reader! I had intended to address thee, reader, with a longer preface. Yet the mass of political affairs which keep me more than usually busy these days, and the hasty departure of our Kepler, who intends to leave for Frankfurt within the hour, only left me a moment's time to write. But I thought nevertheless that I ought to address a few words to thee, lest ye should be-

come confused by the liberties which Kepler takes in deviating from Brahe in some of his expositions, particularly those of a physical nature. Such liberties can be found in all philosophers since the world existed; and it in no way affects the work of the Rudolphine tables. [This refers to the planetary tables dedicated to Rudolph which Tengnagel had promised to produce, and never did.] You will be able to see from this work that it has been built on the foundations of Brahe . . . and that the entire material (I mean the observations) was collected by Brahe. In the meantime, consider Kepler's excellent work . . . as a prelude to the tables and to the observations to follow which, for the reasons explained, can only be published slowly. Pray with me to the Almighty and all-wise Lord for the rapid progress of this much-desired work and for happier days.

> Franz Gansneb Tengnagel,
> In Campp. Counselor of his Imperial Majesty."

4 G.W., Vol. XV, p. 131 *seq.*
5 D. Fabricius to Kepler, 20.1.1607, G.W., Vol. XV, p. 376 *seq.*
6 30.10.1607, G.W., Vol. XVI, p. 71.
7 The writer is the Danzig astronomer P. Crueger, quoted in W. v. Dyck and M. Caspar, *Nova Kepleriana* 4, Abhandlungen der Bayrischen Ak. d. Wiss. XXXI, p. 105 *seq.*
8 Loc. cit.
9 *Astronomiae Pars Optica*, Dedication to Rudolph II, G.W., Vol. II.
10 Letter to Besold, 18.6.1607, G.W., Vol. XV, p. 492.
10a Letter to Herwart, 10.12.1604, G.W., Vol. XV, p. 68 f.
11 Letter to Herwart, 24.11.1607, G.W., Vol. XVI, p. 78 *seq.*
12 Letter to D. Fabricius, 11.10.1605, G.W., Vol. XV, p. 240 *seq.*
13 *Dissertatio cum Nuncio Sidero*, G.W., Vol. IV, p. 281 *seq.*
14 There has been some controversy on the question whether the title meant "messenger" or "message"—cf. Stillman Drake, *Discoveries and Opinions of Galileo*, New York, 1957, p. 19. Stillman Drake translates the title as *The*

Starry Messenger; de Santillana (see note 27a, Chap. 8), as *Sidereal Message* (*Dialogue*) or *Starry Message* (*The Crime of Galileo*). I propose to use *Messenger from the Stars,* or *Star Messenger* for short.

Chapter Eight   KEPLER AND GALILEO

1   F. Sherwood Taylor, *Galileo and the Freedom of Thought,* London, 1938, p. 1.
2   This is strictly true for small angles only, but sufficient for practical purposes of time measurement. The correct law of the pendulum was discovered by Huygens.
   The candelabra still shown at the Cathedral of Pisa, whose oscillations are alleged to have given Galileo his idea, was only installed several years after the discovery.
3   His manuscript treatise *De Motu,* written about 1590, and privately circulated, certainly deviates from Aristotelian physics, but by subscribing to the entirely respectable theory of impetus which had been taught by the Paris school in the fifteenth century and by several of Galileo's predecessors and contemporaries. Cf. A. Koyré, *Etudes Galileennes,* Paris, 1939.
4   About his technical treatise on the proportional compass, see note 12.
5   Letter to Maestlin, September 1597, G.W., Vol. XIII, p. 140 *seq.*
6   G.W., Vol. XIII, p. 130 f.
6a   *Trattato della Sfera, Opere, Ristampa della Ediz. Nazionale,* Florence, 1929–39, Vol. II, pp. 203–255. Henceforth *Opere* refers to this edition, except when marked "ed. F. Flora," which refers to the handier selection of works and letters in one volume, published in 1953.
7   Quoted by Sherwood Taylor, op. cit., p. 85.
8   G.W., Vol. XIII, p. 144 *seq.*
9   G.W., Vol. XIV, p. 256.
10   Ibid., p. 441.
11   Ibid., p. 444 f.
12   It is surprising to read that Professor Charles Singer attributes the discovery that the nova of 1604 had no parallax to

Galileo, and, moreover, passing in silence over Tycho's classic book on the nova of 1572, writes:

"New stars when previously noticed had been considered to belong to the lower and less perfect regions near the earth. Galileo had thus attacked the incorruptible and interchangeable heavens and had delivered a blow to the Aristotelian scheme, wellnigh as serious as the experiment on the tower of Pisa (sic)." (Charles Singer, *A Short History of Science to the Nineteenth Century*, Oxford, 1941, p. 206.)

Since that experiment is also legendary, Professor Singer's comparison contains an ironic truth; but this triple misstatement is characteristic of the power of the Galileo myth over some eminent historians of science. Professor Singer also seems to believe that Galileo invented the telescope (op. cit., 217), that in Tycho's system "the sun revolves round the earth in twenty-four hours carrying all the planets with it" (ibid., p. 183), that Kepler's Third Law was "enunciated in the *Epitome Astronomiae*" (ibid., p. 205), etc.

13 Quoted by Ernst Zinner, *Entstehung und Ausbreitung der Copernicanischen Lehre* (Erlangen, 1943), p. 514.

14 *Le Operazioni delle Compasso Geometrico e Militare*, Padova, 1606; *Opere* II, pp. 362–405.

15 *Usus et Fabrica Circui Cuiusdam Proporziones*, Padova, 1607; *Opere* II, pp. 425–511.

16 Capra's teacher was the distinguished astronomer Simon Marius (1573–1624), discoverer of the Andromeda Nebula, with whom Galileo later became involved in another priority quarrel.

17 Letter to B. Landucci, quoted by Gebler, *Galileo Galilei and the Roman Curia*, London, 1879, p. 19.

18 George Fugger (a member of the famous banker's family) in a letter to Kepler, 16.4.1610, G.W., Vol. XVI, p. 302.

18a Cf. Zinner, op. cit., p. 345 f.

19 This refers to the first, Latin edition.

20 *Paradise Lost*, Book ii, l. 890.

20a *Peregrinatio contra Nuncium Sydereum*, Mantua, 1610.

21 *Ignatius his Conclave.*

22 *Opere*, ed. F. Flora, Milano–Napoli, 1953, p. 887 *seq.*

23 Ibid., p. 894 *seq.*
24 28.5.1610, G.W., Vol. XVI, p. 314.
25 Quoted by E. Rosen, *The Naming of the Telescope,* New York, 1947.
26 Letter to Horky, 9.8.1610, G.W., Vol. XVI, p. 323.
26a "Poor Kepler is unable to stem the feeling against Your Excellency, for Magini has written three letters, which were confirmed by 24 learned men from Bologna, to give effect that they had been present when you tried to demonstrate your discoveries . . . but failed to see what you pretended to show them." M. Hasdale to Galileo, 15.4 and 28.4. 1610, G.W., Vol. XVI, pp. 300 f, 308.
27 G.W., Vol. XVI, p. 319 *seq.*
27a It was probably this letter which lead Professor de Santillana to the erroneous statement: "It took even Kepler, always generous and open-minded, a whole five months before rallying to the cause of the telescope. . . . His first *Dissertatio cum Nuncio sidereo,* of April, 1610, is full of reservations." (*Dialogue on the Great World Systems,* Chicago, 1937, p. 98 n.) Kepler's reservations referred, as we saw, to the priority of the invention of the telescope, not to Galileo's discoveries with it.
28 G.W., Vol. XVI, p. 327 *seq.*
29 Except for a short note of introduction to Kepler, which Galileo gave a traveler seventeen years later, in 1627. *Opere* XIII, p. 374 f.
30 Gebler, op. cit., p. 24.
31 At least, that seems to be the meaning. The word *"umbistineum"* does not exist and may either be derived from *"ambustus,"* burned up, or *"umbo"*=boss, projection.
32 9.1.1611, G.W., Vol. XVI, p. 356 *seq.*
33 *Narratio de Observatis a se quatuor Iovis sattelitibus erronibus.*
34 25.10.1610, G.W., Vol. XVI, p. 341.

Chapter Nine CHAOS AND HARMONY

1 The book should really be called "Dioptrics and Catoptrics," for it deals with both refraction and reflection.

2 Except for the preface.
3 *Ad Vitellionem Paralipomena, quibus Astronomiae Pars Optica.*
4 3.4.1611, G.W., Vol. XVI, p. 373 *seq.*
5 Dedication of the *Eclogae Chronicae*, 13.4.1612, quoted in *Johannes Kepler in seinen Briefen*, Vol. I, p. 391 *seq.*
6 Ca., p. 243.
7 Ca., p. 252 f.
8 Ca., p. 300.
9 Galileo was submitted to the much milder form of *territio verbalis*, without actually being led into the torture chamber.
10 Quoted in *Johannes Kepler in seinen Briefen*, Vol. II, p. 183 f.
11 *Harmonices Mundi Libri V*, Linz, 1619. The work is sometimes erroneously referred to as "Harmonices," as if the "s" stood for the plural, whereas it stands, of course, for the genitive.
12 Kepler's translation of the word is *unwissbar*.
13 *Harmonice Mundi*, Bk. V, Cap. 4.
14 Loc. cit.
15 Ibid., Cap. 7.
16 Dedication of the *Ephemerides* for 1620 to Lord Napier.
17 Ibid.
18 *Harmonice Mundi*, introduction to Book V.
19 "*Sed res est certissima exactissimaque, quod proportio, quae est inter binorum quorumconque planetarum tempora periodica, sit praecise sesquialtera proportionis mediarum distantiarum, id est orbium ipsorum.*" (Ibid., V, Cap. 3, Proposition No. 8.)
20 Loc. cit.
21 Loc. cit.
22 Ibid., Appendix to Book V.

Chapter Ten COMPUTING A BRIDE

1 G.W., Vol. XVII, p. 79 *seq.* This is a compressed version.

Chapter Eleven THE LAST YEARS

1 To Bianchi, 17.2.1619, G.W., Vol. XVII, p. 321 *seq.*
2 To Bernegger, 20.5.1624, *Johannes Kepler in seinen Briefen*, Vol. II, p. 205 f.
3 1.10.1626. Ibid., Vol. II, p. 222 ff.
4 Loc. cit.
5 Kepler became acquainted with Napier's logarithms in 1617: "A Scottish baron has appeared on the scene (his name I have forgotten) who has done an excellent thing by transforming all multiplication and division into addition and substraction. . . ." (Ibid., Vol. II, p. 101.) Since Napier did not at first explain the principle behind it, the thing looked like black magic and was received with skepticism. Old Maestlin remarked: "It is not fitting for a professor of mathematics to manifest childish joy just because reckoning is made easier." (Ca., p. 368.)
6 To Bernegger, 20.5.1624, see note 2.
7 Ca., p. 302.
8 To Bernegger, 6.4.1627, *Johannes Kepler in seinen Briefen*, Vol. II, p. 236 *seq.*
9 To Bernegger, 22.7.1629, ibid., p. 292.
10 To Bernegger, 2.3.1629, ibid., p. 284 f.
11 To Bernegger, 29.4.1629, ibid., p. 286 f.
12 To Philip Mueller, 27.10.1629, ibid., p. 297.
13 Cf. Marjorie Nicolson's essay, "Kepler, the Somnium, and John Donne," in her *Science and Imagination*, Oxford, 1956.
14 Letter to Bartsch, 6.11.1629, *Johannes Kepler in seinen Briefen*, Vol. II, p. 303.
15 To Philip Mueller, 22.4.1630, ibid., p. 316.
16 Bartsch to Philip Mueller, 3.1.1631, ibid., Vol. II, p. 329.
17 Ibid., Vol. II, p. 325.
18 Ca., p. 431.
19 Quoted by S. Lansius to anon., 24.1.1631, *Johannes Kepler in seinen Briefen*, Vol. II, p. 333.

20 From which expression one might infer that they refused him the last sacraments.

21 To Bartsch, *Johannes Kepler in seinen Briefen*, Vol. II, p. 308.

INDEX

Index